FACTORS CONTROLLING THE BEHAVIOR
OF TIBIAL SHAFT FRACTURE

A CORRELATION OF LABORATORY AND CLINICAL STUDIES

AUGUSTO SARMIENTO MD

LOREN L. LATTA, P.E.

DEPARTMENT F ORTHOPAEDICS AND REHABILITATION, UNIVERSITY OF MIAMI

SCHOOL OF MEDICINE, MIAMI, FLORIDA 33152

Comments about this paper.

*This study was done and submitted to The Kappa Delta Sorority and the Orthopaedic Research and Education Foundation of the American Academy of Orthopaedic Surgeons. It was awarded the **Kappa Delta Award** in 1976.*

Kappa Delta awards honor innovative orthopaedic research .

The Kappa Delta Sorority and the Orthopaedic Research and Education Foundation (OREF) presents awards to scientists conducting outstanding research related directly to musculoskeletal disease or injury with the goal of advancing patient treatment and care.

Published May 2018

ISBN: 978-1-387-81227-1

Index

Introduction

Fracture of the tibial shaft have long attracted the interest of orthopaedist as well as other scientists and investigators because of difficult management problems, the high incidence of non-union, the devastating consequences of osteomyelitis, and the long period of incapacitation of suffered during healing. Perhaps more than any other fracture of the long bones of the appendicular skeleton, the management of the tibial fracture has been subjected to the principal of immobilization of adjacent of the fragments, the better the ultimate result. Robert Jones' teaching that "rest, enforced, uninterrupted, and prolonged, is the main foundation of sound healing" has been applied to tibial fractures with great emphasis

The closed functional method of treatment for fractures of the tibia developed at our institution, points to early joint mobilization and weight bearing ambulation [50, 52, 53, 55] The results obtained with the method have indicated that neither immobilization of the joints and//or fragments are prerequisites for fracture healing. We believe that early function of joints and muscles, and the graduated subjection of weight bearing loads favorably influences osteogenesis. Our clinical experience, support by animal investigations, have strongly indicated that motion at the fracture site, if cause by functional activities, is beneficial rather than detrimental to fracture healing.

In order to obtain valid, scientific explanations for the apparent success with this method of treatment, we conducted animal and biomechanical studies. In the following report we will describe our animal investigations or fracture healing in regard to the influence played by function in the mechanical properties of callus. We will also present studies dealing with the role of the soft tissues in determining initial and subsequent behavior in the fractures; in vivo, cineradiographic analysis of fracture behavior under various loading conditions in comparison to the laboratory models; and a summary of our clinical experiences with 532 cases of tibial fractures which we treated with functional cast and braces that permit mobilization of all joints, unencumbered use of the musculature of the extremities, and early weight-bearing ambulation.

Section I

ANIMAL LABORATORY STUDIES ON FRACTURE HEALING

ANIMAL LABORATORY STUDIES ON FRACTURE HEALING:

The Effects of Function in Experimentally Created Fractures in Rats

There has been little experimental work which correlates function with fracture healing on a microscopic and mechanical basis. We designed the following study specifically to assess the influence that function may have on fracture healing. In order to accomplish this, a careful examination of the fracture callus at regular time intervals during the healing process was carried out. Differences in the fracture callus of immobilized and functional animals were based on x-ray, microscopic and mechanical findings.

Materials and Methods

One hundred and seventy-nine adult male Sprague-Dawley rats were used for this experiment. Animals were obtained from a supply house to assure uniformity of breed. Weights averaged between 275 and 325 grams. Care was taken to select rats whose weights were in the 300 gram range since this weight group proved to be the easiest to work with experimentally. The animals were maintained on standard laboratory feed and water. In order to assure greater freedom of movement only two animals were kept in each cage.

The femur was chosen for the experimental fracture because of its uniform tubular shape and easy accessibility. In order to create a standard experimental fracture which could be consistently reproduced with the same alignment and degree of displacement so that bending tests could be used for mechanical assessment, a modification of the method of Jackson and Reed was used. It consisted of the introduction of an intramedullary pin, 0.035 mm in diameter and 2.5 cm in length, via a retrograde-manner percutaneously through the distal femur prior to fracture.

Actual fracturing was done using a pneumatic device with a rounded wedge bar fixed to the end of a piston (Fig. l). This piston was powered by a cylinder of compressed air and controlled by a simple pressure gauge and valve. The wedge had a total excursion of 4mm in order to immobilize the joints above and below the fracture site in one group, a Zorac spica casts was fashioned. X-rays were taken immediately following the fracturing and casting and at weekly intervals throughout the experiment. Fractures which were not mid-shaft or showed comminution were not used in this study. Animals were sacrificed as needed using an overdose of ether.

Following sacrifice, each fractured femur was removed as a unit. Extreme care was taken not to traumatize the healing fracture site and callus during removal of all soft tissue about the femur. In actuality, the muscles adjacent to the callus were easily removed due to the fact that a periosteal seal rapidly formed over the developing callus indicating a rapid healing process [35] The many microscopic vascular connections between the callus and soft tissue were disrupted through preparation. The pin, which was initially required to create the standard fractures and prevent significant displacement, was easily extracted with a hemostat through the point of insertion. Specimens for histologic analysis were fixed in formalin, decalcified and blocked in paraffin for sectioning. Sagittal and cross sectional cuts were made and stained with hematoxylin and eosin. Specimens, which would undergo mechanical testing, were kept moist in physiologic saline solution and refrigerated. All mechanical tests were used to determine the strength of the healing fracture and the modulus of elasticity (material stiffness). To do this, approximate calculations were made of the bending stresses, which developed at the fracture site, under loading. In the formula $O = M_C/I_X$, O equals the maximum tensile or compressive stress due to bending; M equals the bending moment; c equals the distance from the neutral axis to the outermost fibers; and IX equals the area moment of inertia. The area moment of inertia was calculated by approximating the cross section at the healing fracture as an oval. Measurements for calculations of IX were made with Vernier calipers for major and minor diameters on each fracture site tested.

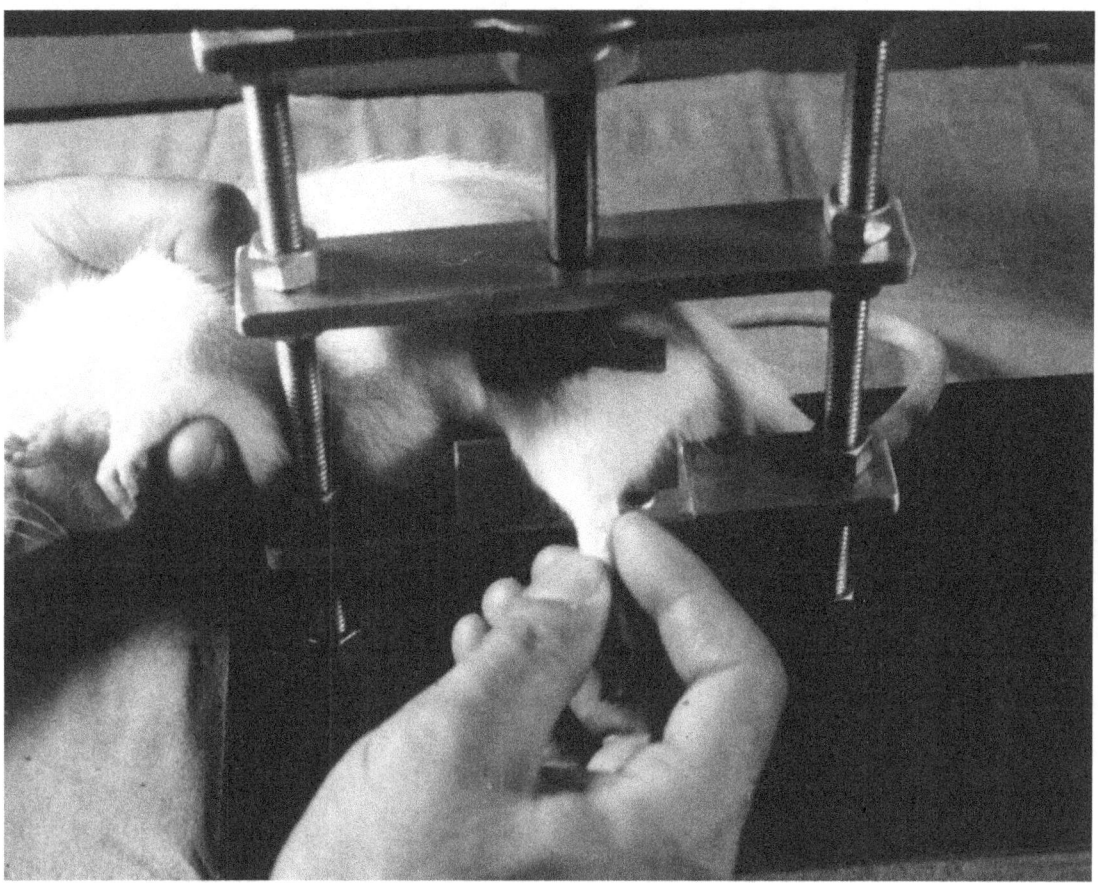

Figure 1 – Pneumatic fixture for fracturing rats' femur by bending.

Actual testing was done on an apparatus consisting of a load cell and deflection transducer which were secured into the compression cage of a Dillon Testing Machine (Fig. 2).Load was plotted on the Y coordinate and deflection on the X coordinate of the recorder.

In brief, the entire fracture specimen was supported on two parallel bars, 0.90 inches apart, with the fracture callus midway between them. The load cell rested on the anterior surface of the fracture callus while the deflection transducer rested directly beneath on the posterior surface. All specimens were supported and loaded in this manner for uniformity. Deflection was at a constant rate of 0.007 inches per second.

Calculations were made for load stress and modulus of elasticity. Significant levels were obtained using the Student's T Test. 'Per cent difference between groups were also recorded.

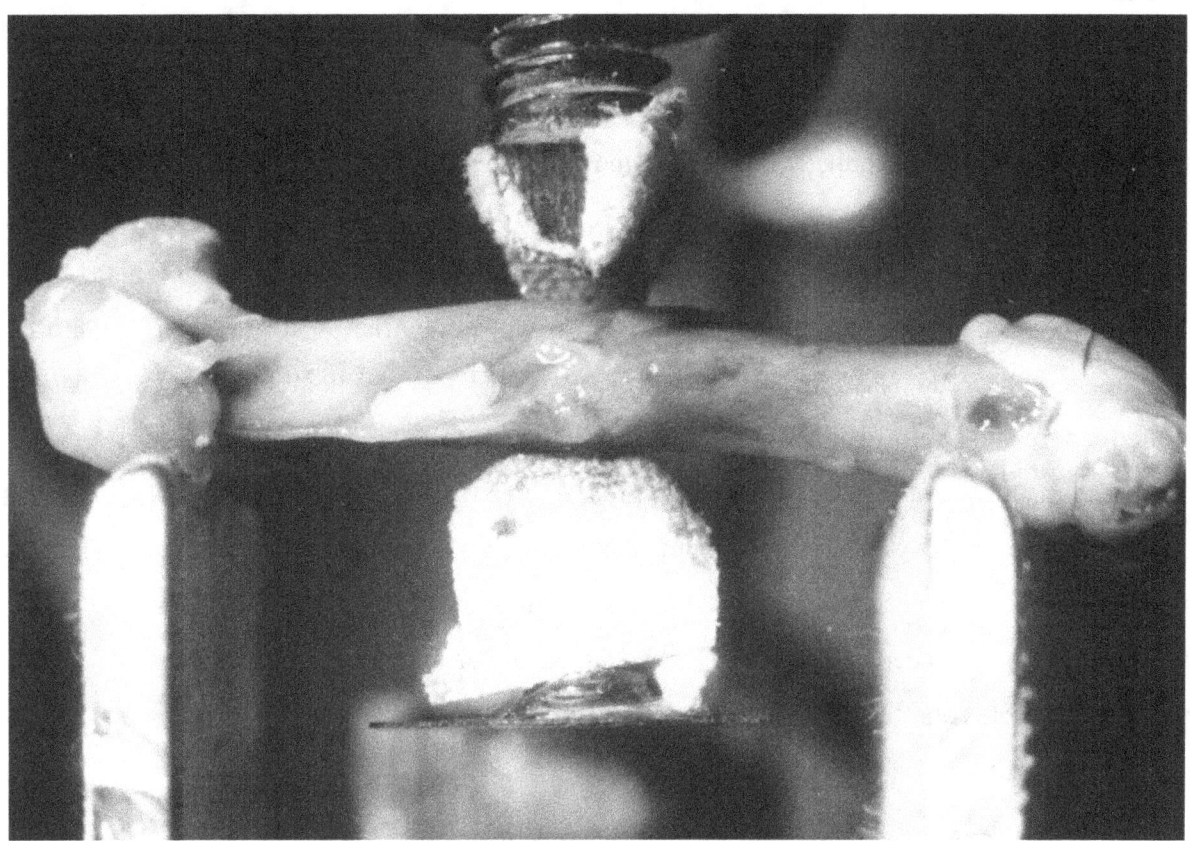

Figure 2 – The bending test of a rat's femur.

Results

Histologic slide preparations of the fracture sites and callus formed in both immobilized and functional groups were made throughout the experiment.

The initial response to the fracture in both groups was essentially the same. This included the formation of a small fracture hematoma, an acute inflammatory response and the appearance of necrotic marrow debris at the fracture site. There was notably little hemorrhage at the fracture site of all specimens examined immediately following the fracture. At all times, the vascular supply of the cortex adjacent to the fracture site showed necrosis. In both, periosteal hypertrophy was present for some distance away from the fracture site itself. The cambium layer of the periosteum showed increased mitotic activity. Periosteal elevation became more pronounced as osteogenic cells and loosely arranged fibrous bone formed between the cortex and the periosteoum.

However, by the end of the first week qualitative differences in the callus formed were apparent. Cartilage production in the weight bearing group was conspicuously more prevalent (Fig. 3). This gave the fracture callus in this group a massive appearance. (Cartilage formation is typical of fracture healing in this research animal and was seen in both groups). Likewise, new bone formation occurring directly from the osteogenic cells of the periosteum was more prominent in the weight bearing group than in the immobilized group. The callus continues to grow and the differences were even more pronounced histologically towards the end of the third week of fracture healing (Fig. 4). These differences were in the degree

Fig. 3 Photomicrographs depicting the changes the fracture site of an ambulatory animal, at 2 weeks. The external callus consists of subperiosteal new bone and a central mass of cartilage adjacent the fracture site. Few internal or emedullary changes have occurred. (H & E; x 7.5)

Fig. 4 Cross sectional photomicrographs at 3 weeks of a fracture specimen from the ambulatory group. Exuberant periosteal new bone in present about the original cortex. (H & E: x 7.5)

of response to the many factors associated with functional ambulation rather than differences in the type of response. This point, which is of considerable importance both clinically and experimentally, will be given more consideration later in the article.

At three weeks the large external callus is still present in the functional group. In addition abundant new bone is present subperiosteally along the adjacent diaphyseal cortex as well as at the fracture site. Much of the cartilaginous portion previously mentioned has already been replaced by new bone by the process of enchondral ossification through vascular invasion and callus shrinkage begins (Fig. 5). This stage of the process corresponds to the strength and corresponding callus size changes observed in similar studies by Lindsay. [133A] (The external callus is dependent on the adjacent soft tissues for its blood supply and many cross connections exist between the vascular network of the callus and soft tissue-

Fig. 5 Photomicrographs depicting the changes at fracture site of an ambulatory animal at 2 weeks.

Rapid replacement of the cartilage is occurring through the process of enchondral ossification. Note he many vascular buds present that appear to be destroying or eroding away the cartilaginous portion of the callus. (H & E: X 200)

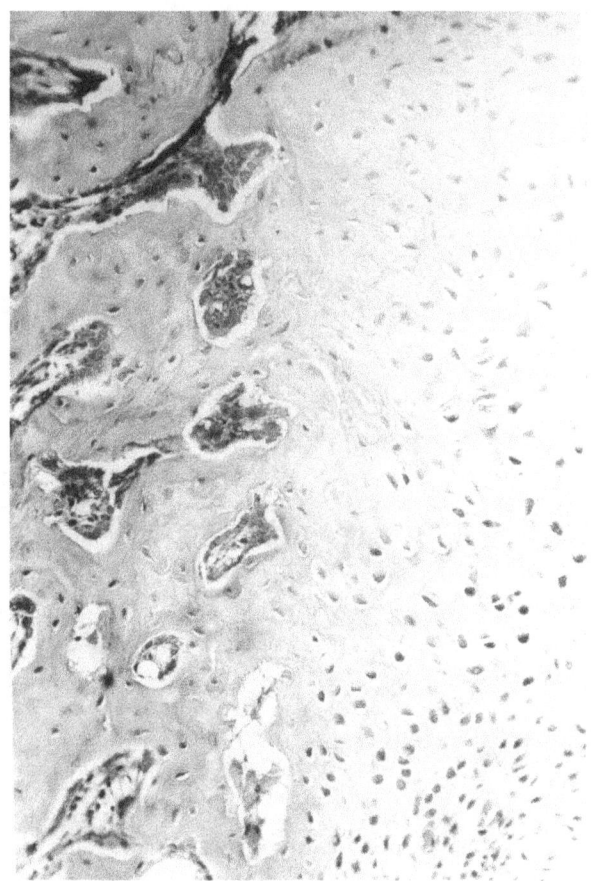

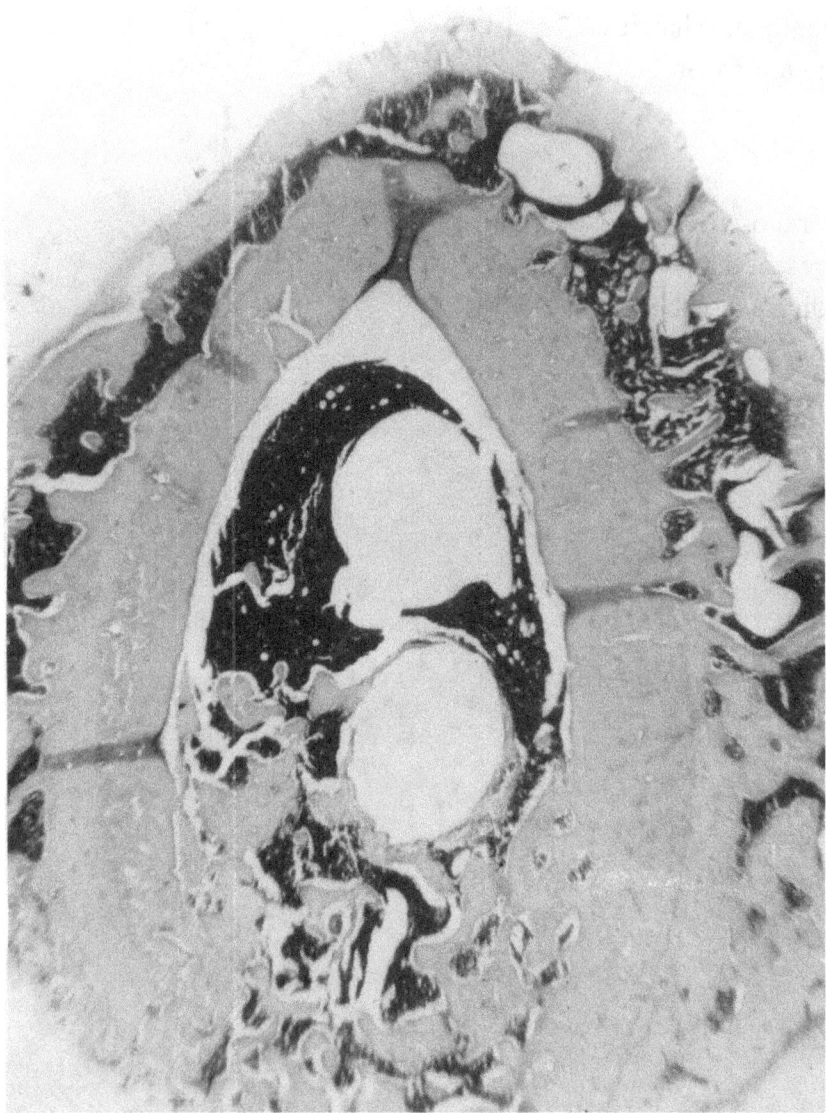

Fig. 6 Cross sectional photomicrographs at 3 weeks of a fracture specimen from the immobilized group.

Periosteal new bone formation is present but to a lesser degree than in Fig 4. The fine trabecular pattern is also lacking.

(H & E: X 7.5)

Through the process of remodeling the internal callus is rapidly replaced by a normal medullary canal and narrow elements

By the fourth week, remodeling and subsequent shrinkage of the callus is the prominent feature in fracture healing. The many fine trabeculae of a new bone seen in the early callus of the weight-bearing group have now been replaced by broader and more compact ones. In Fact there are many points of fusion and actual invasion of the original cortex by the new bone compared with 3 weeks. (Figs. 8 & 9)

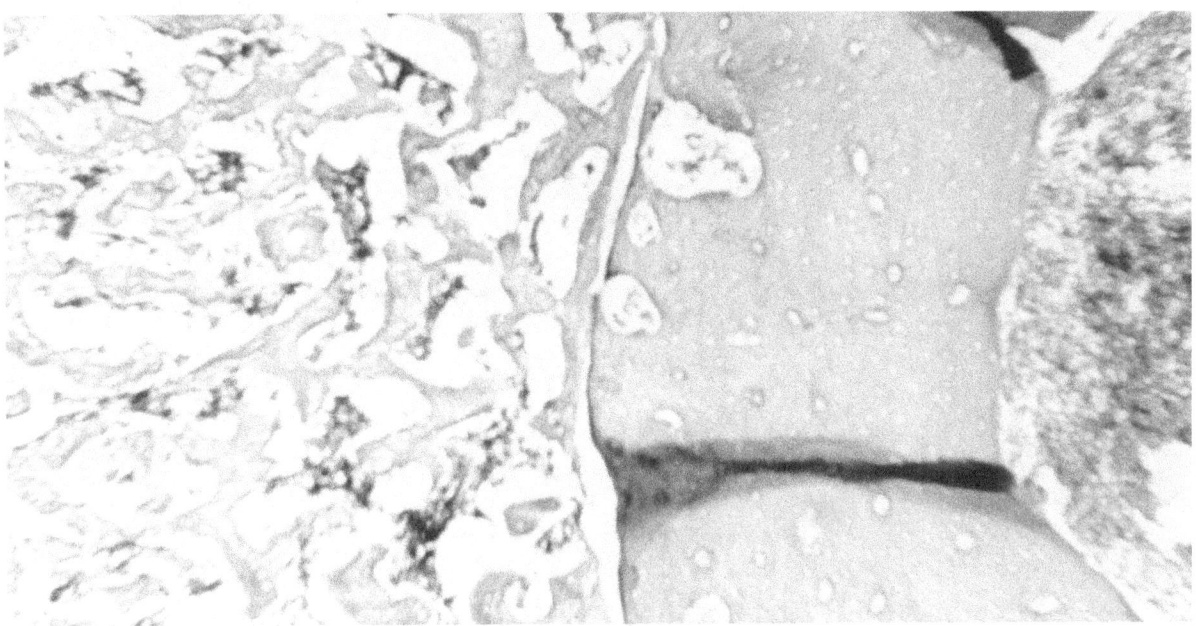

Fig. 8   Higher power view revealing the fine, lacy trabecular pattern of the new bone. Large cortical vascular are present. Large cortical vascular channels are present. The preparation has left many empty lacuna which would normally contain osteocytes. (H & E; X 40)

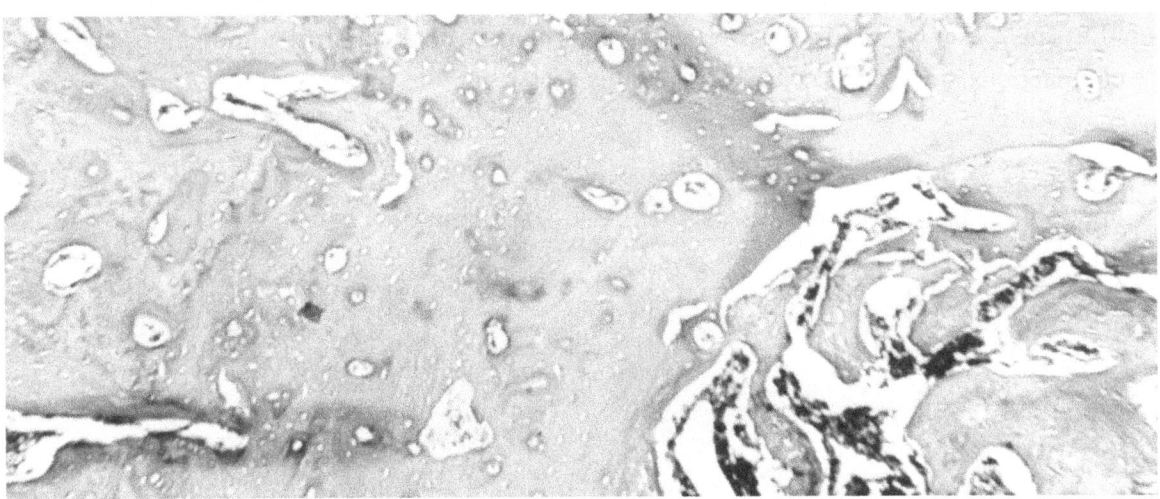

Fig. 9   An enlargement showing blending of the new bone with the original cortex. Many vascular cross connections are present. This blending or remodeling of new bone and cortex provides increased structural stability. (H & E ; X 40)

This new bone about the original cortex takes on the physical characteristics of the outer circumferential lamellae. A well-formed vascular network is now present throughout the new bone and original cortex with many fine connections existing between the new bone and the original cortex. (Fig. 10)

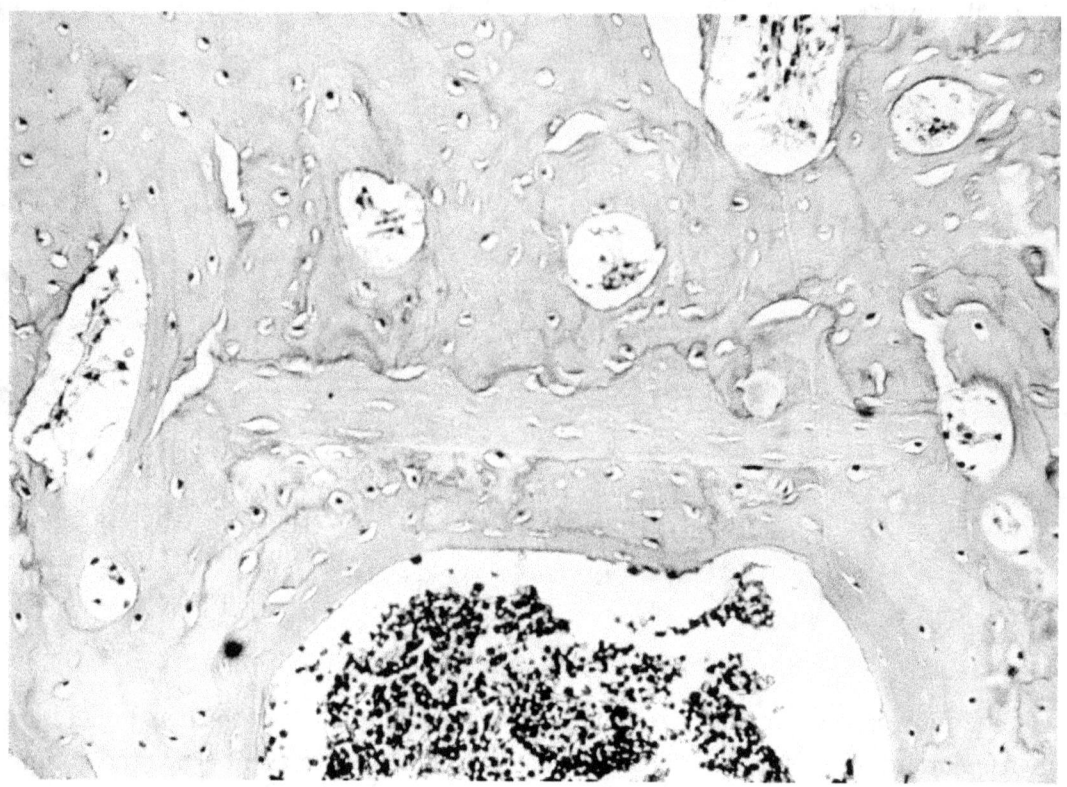

Fig. 10 Cross sectional photomicrographs at 4 weeks of fracture specimen in ambulatory group. Reveals a highly organized external callus. The new bone is fairly homogenous and more compact than specimens of earlier fractures. (H & E ; X 7.5)

In the immobilized group, at four weeks, there is less apparent new bone and vascular connections. (Figs. 11 & 12)

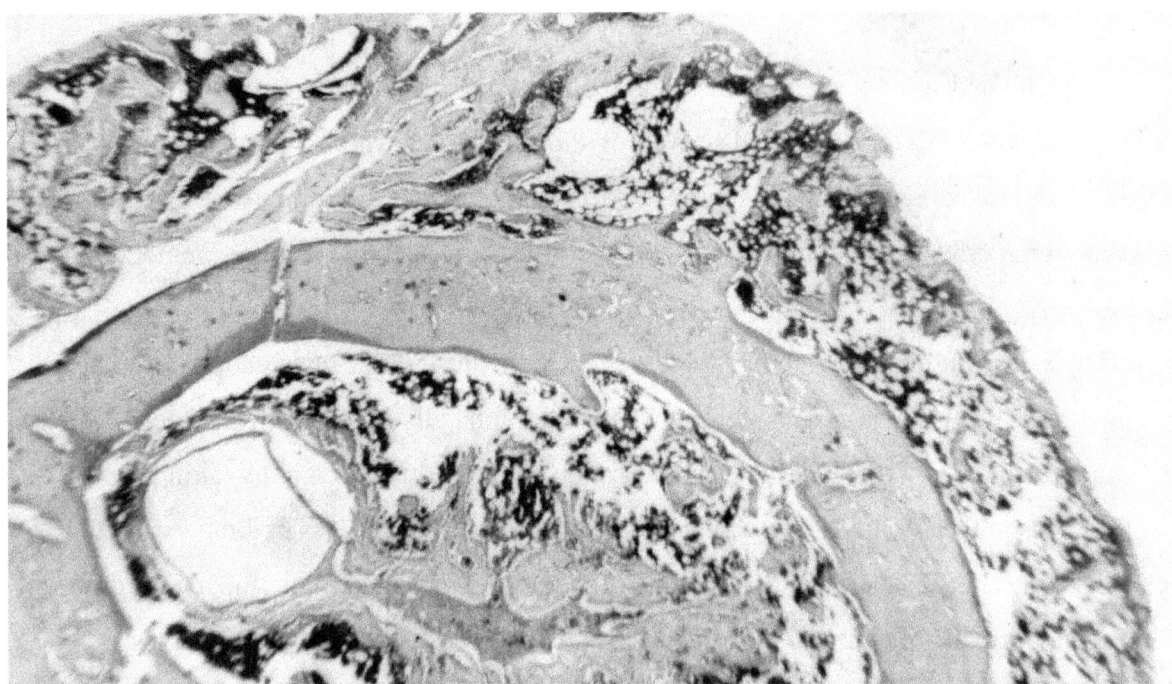

Fig. 11 Cross sectional photomicrographs of fracture specimens at 4 weeks in the immobilized group. Fracture site to resemble the pre-fracture from through remodeling. (H & E; x 7.5)

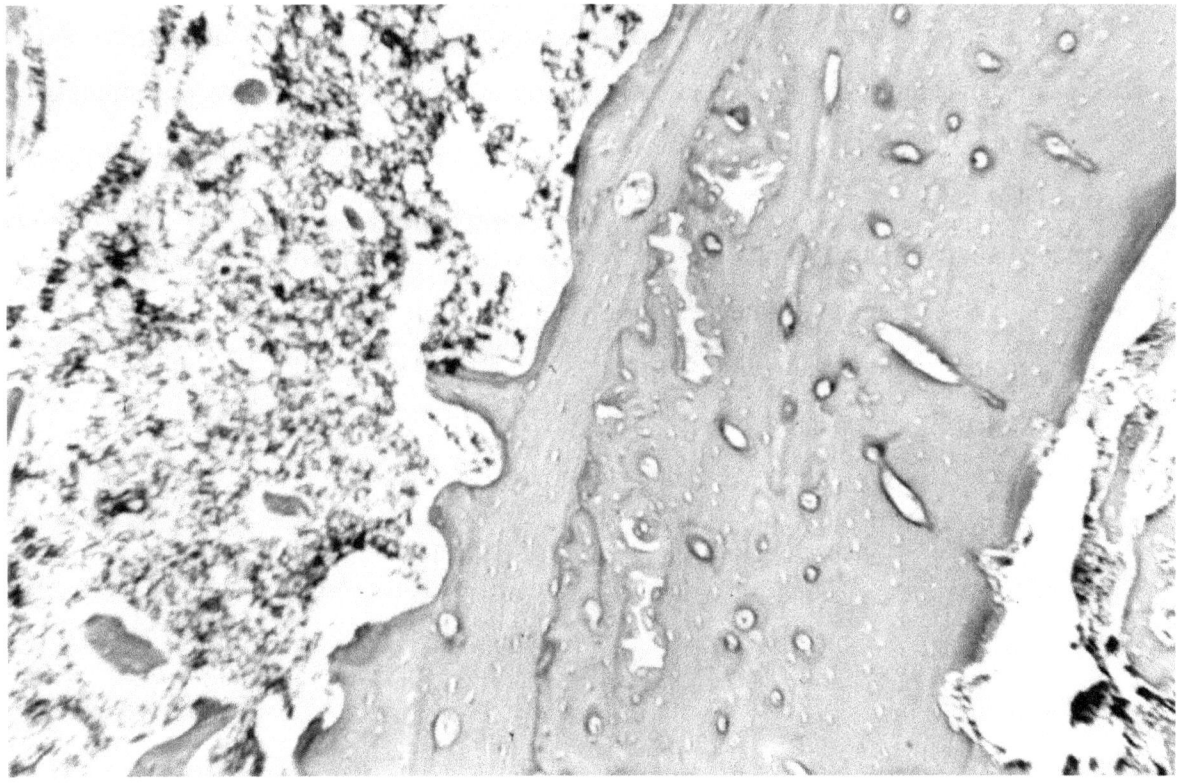

Fig. 12 Higher power view. Some endosteal callus is present. (H & E; x 40)

The fifth and sixth week periods reveal the final stages of fracture repair from the histologic viewpoint. Remodeling had slowed. In those fractures where immobilization was carried out, the fracture sites have in remodeled to resemble the pre-fracture state appearance. In the functional group cortical hypertrophy (thickening) has taken place during remodeling.

The x-ray findings were generally consistent with those made microscopically. The large fracture callus in the functional group was radiographically apparent by the second week of fracture healing due to the fact that mineralization of the callus was rapidly taking place. X-ray differences in the size of the fracture callus were still present but less striking by the third week (Fig. 13). The presence of a radiographic fracture line persisted in the functional group well into the third and fourth weeks of fracture healing. This was not the finding in the immobilized group in which the fracture line had uniformly disappeared by the end of the fourth week.

Although the appearance of the fracture line on x-ray suggests that healing is not complete, its disappearance did not mean a strong union was present. [69] Stability and strength cannot be determined by x-ray in absolute terms. For this reason a healed fracture, for this experiment, is
one that is capable of withstanding the same mechanical forces as a non-fractured specimen. This definition is solely made so that an end point can be established and comparisons made. "Clinical union"and "radiographic union "are terms that are not synonymous with this definition of a healed fracture and will not be used in this article. In order to establish control values, intact non-fractured femurs were subjected to the same mechanical tests.

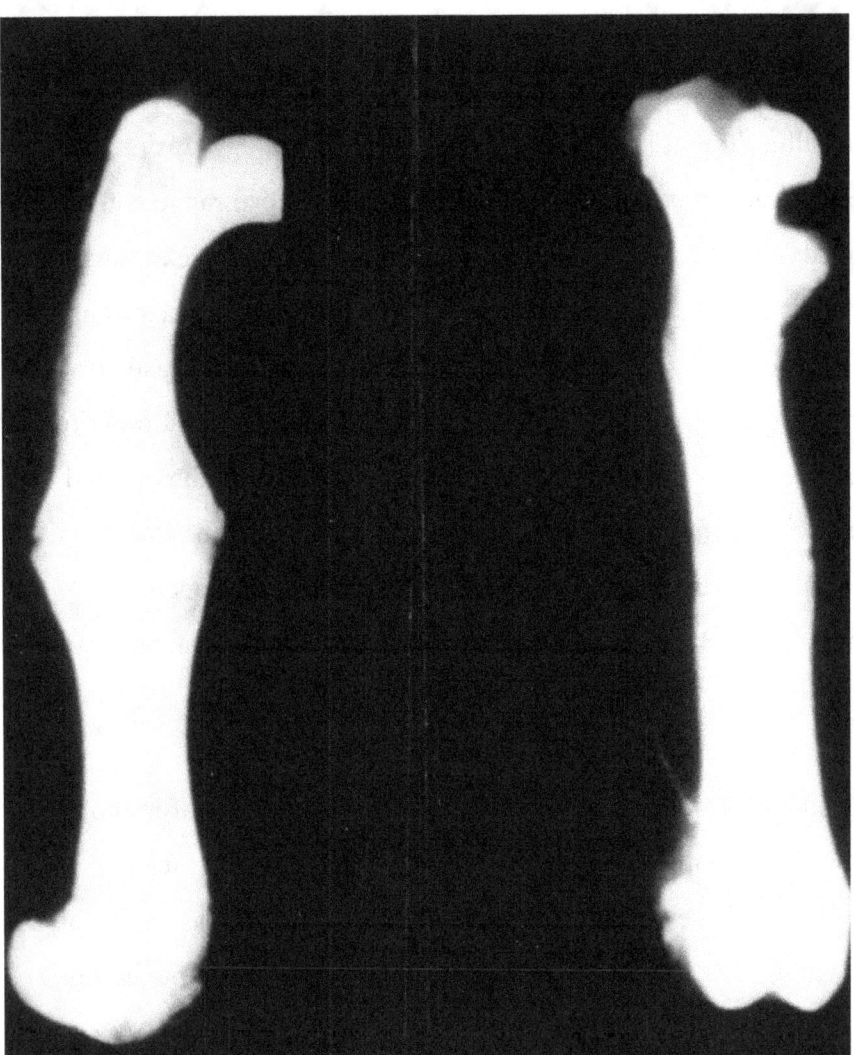

Fig. 13 Radiograph of 3 weeks old femoral fracture from the ambulatory group. Note the fusiform shape of the healing bone at the fracture site.

(A).   Radiograph of 3 weeks old femoral fracture from the immobilized group. Less architectural changes have taken place about the fracture site (B).

The first physical parameter to be described and compared in both groups is the loading characteristics of the fracture callus. 'Initial loading of the specimens in both groups reveals that the callus has a significant elastic range (assuming that the linear proportional range is the elastic range). However, as loads are increased, this linear relationship no longer exists and permanent deformation at the fracture site occurs. Deformation in which the loaded specimen fails to return to its original state following removal of an applied load is termed plastic or permanent deformation. Control specimens also exhibit significant plastic deformation prior to failure or fracture. As healing progresses the load required to produce failure at the original fracture site increases. 'Likewise, the amount of displacement which occurs at the fracture site prior to failure is dependent on the degree of healing. with maturation less permanent displacement occurs prior to fracture. (By the fourth week both cast and uncast fractures were relatively brittle).

Comparison of the mean load sustaining abilities of the healing fractures of immobilized and functional animals reveals significant levels of difference. At two weeks the mean maximum load sustaining ability for those fractures in the immobilized group was 76.7% of that sustained by the fractures in the functional group. By three weeks, the mean maximum group was 52.4% of that sustained-by the fractures in the functional group (Table l).

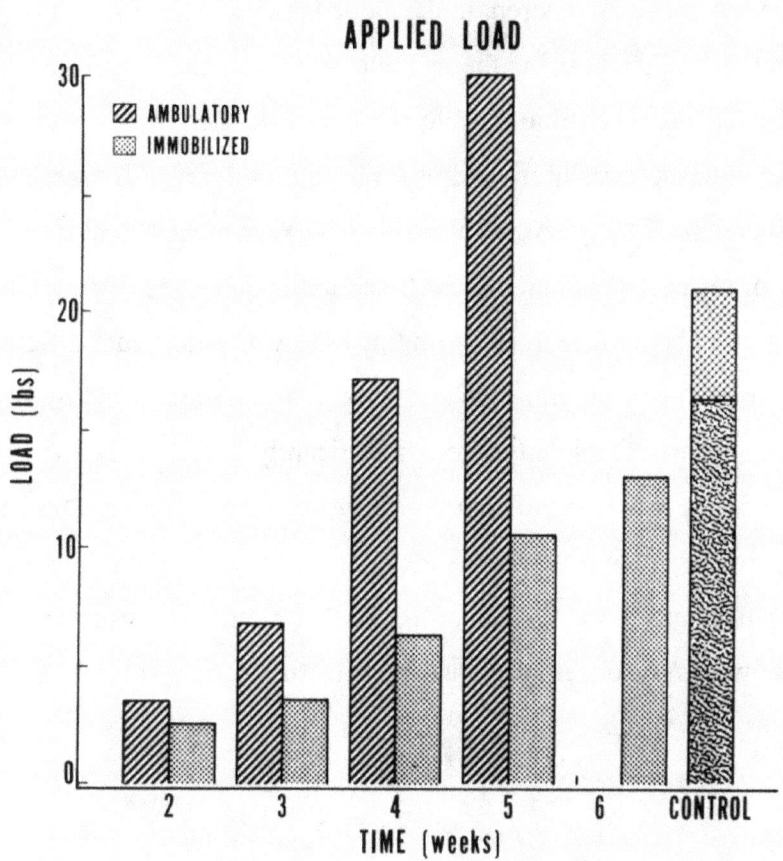

APPLIED LOAD

TABLE I

SUMMARY OF MEAN LOAD VALUES*

Weeks Post Fx.	Immobilized		Functional	
	Elastic Limit	Ultimate Load	Elastic Limit	Ultimate Load
2	$2.20 \pm .252$**	$2.64 \pm .305$	$2.94 \pm .430$	$3.44 \pm .470$
3	$3.20 \pm .553$	$3.50 \pm .603$	$6.16 \pm .840$	$6.80 \pm .840$
4	$5.85 \pm .716$	$6.25 \pm .749$	16.30 ± 2.79	17.10 ± 2.46
5	9.97 ± 1.82	10.40 ± 1.61	28.00 ± 4.10	30.10 ± 3.70
6	12.70 ± 2.99	13.00 ± 2.86		

* All loads are in pounds force.

** Standard deviations at 90% Confidence, by students' t distribution.

The next parameter is the relative strength developed at the fracture site during healing. Strength is calculated in terms of stress using the formula previously mentioned. Both yield strength and ultimate strength are determined. Yield strength is defined in this study as the maximum stress that can be sustained within the linear proportional range. It characterizes the upper limits of the elastic range of the fracture callus. In this experiment, the ultimate strength reflects the maximum tensile stress than can be developed in the outermost fibers of the callus prior to failure in a three point bending test. »Our findings show no significant differences in either yield or ultimate strength when comparing both groups at the second and third week periods; Ability to deform before complete failure is significantly different. The ability to sustain stress at four and five weeks is significantly greater in those fracture specimens subjected to functional stresses. At four weeks the mean ultimate stress in the fracture specimens of the immobilized group was 52.4% of that of the fracture specimens in the functional group. At five weeks the difference was 44.2% of functional (Table 11).

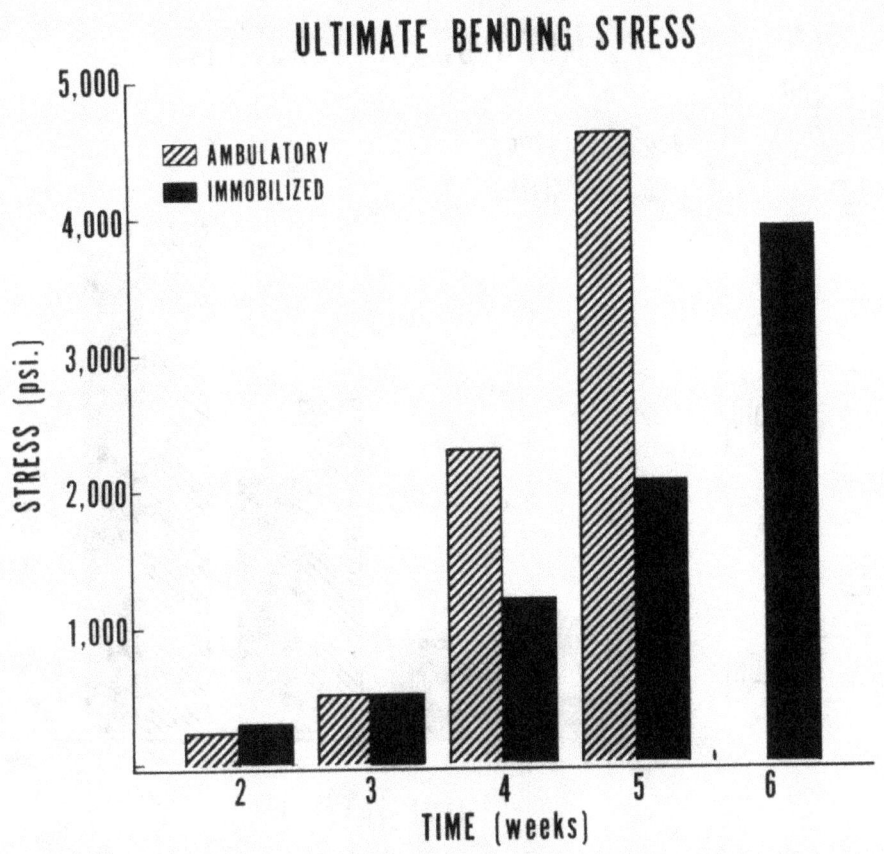

ULTIMATE BENDING STRESS

TABLE II

SUMMARY OF MEAN STRESS VALUES*

| Weeks | Immobilized | | Functional | |
	Yield Stress	Ultimate Stress	Yield Stress	Ultimate Stress
2	$260 \pm 30.7x$	310 ± 37.5	226 ± 32.6	264 ± 32.16
3	480 ± 87.8	530 ± 106.3	505 ± 78.4	556 ± 80.2
4	1140 ± 249	1220 ± 258	2250 ± 679	2330 ± 664
5	1970 ± 311	2060 ± 275	4350 ± 958	4660 ± 972.2
6	3790 ± 917.6	3910 ± 997		

*Stress recorded in pounds per square inch.

x 90% confidence limits as determined by Student's t distribution.

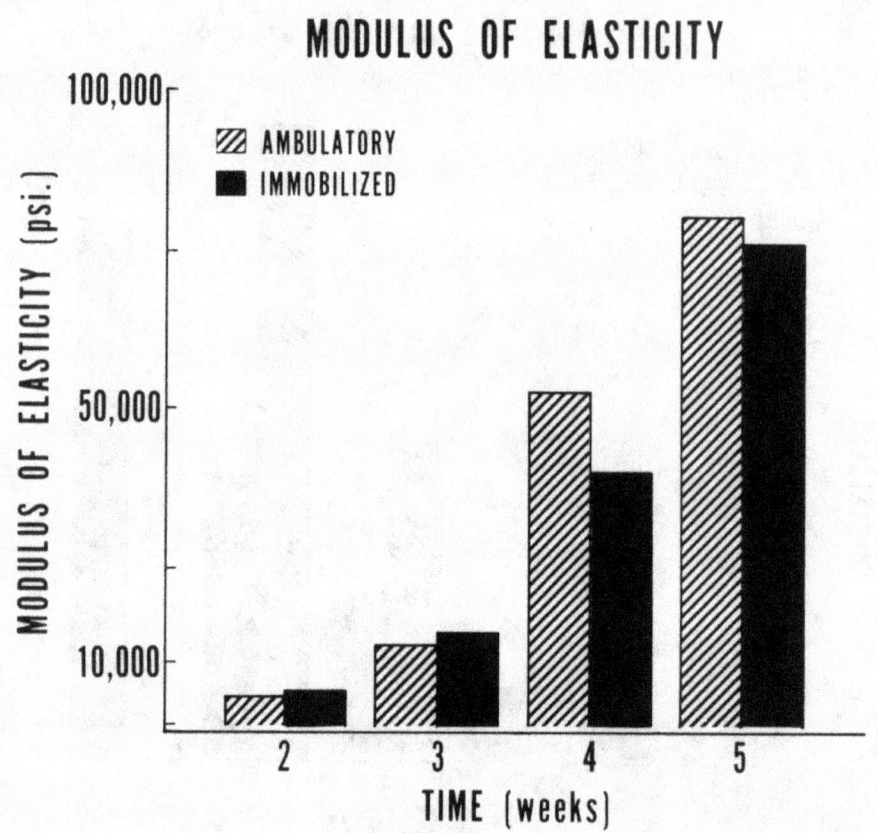

MODULUS OF ELASTICITY

TABLE III

MODULUS OF ELASTICITY

Summary of Mean Values*

Weeks	Immobilized	Functional
2	$5{,}470 \pm 979.7x$	$4{,}670 \pm 862$
3	$19{,}000 \pm 6{,}710$	$13{,}200 \pm 3{,}260$
4	$40{,}200 \pm 9{,}920$	$52{,}800 \pm 15{,}104$
5	$75{,}700 \pm 10{,}600$	$80{,}400 \pm 15{,}311$
6	$190{,}000 \pm 61{,}460$	

* Modulus of Elasticity recorded in pounds per square inch.

x 90% confidence limits as determined by Student's t distribution.

Comparison of callus sites, moments of inertia and flexular stiffness better reflect the effect of these material properties of the whole bone as a structure (Fig. 14).

Discussion

The rate and quality of fracture healing under different physical conditions is difficult to examine experimentally. 'Bone exhibits highly complex mechanical properties.] ,[1,8,44,49] More difficult and still more complex is the callus which forms at the fracture site . [29,64,65] Its organization, development and maturation exists for relatively brief periods of time. Its histologic, radiographic and mechanical properties are in a constant change. At one point in time the fracture callus is a soft pliable mass and at another, a firm highly developed structure capable of providing much stability to the fractured bone. Through the process of healing, the fracture callus loses its particular physical and histologic properties to that of the whole bone. For this reason, this experimental study cannot be all comprehensive. The subject is too vast. However, we feel that both the physical and mechanical properties of healing fractures have been better defined using a testing system which is simplistic and reliable.

Many researchers have already described a relationship between motion at the fracture site and the amount of cartilage formed. [4,32,38,59,72] This proved to be our finding. Cartilage formation occurred in both groups as would be anticipated from the type of experimental animal used. [33,64,65]

Replacement of this cartilage by new bone in the weight-bearing group was rapid and occurred through the process of vascularization and enchondral ossification. Because of its rapid replacement, the formation of cartilage at the fracture site was not considered a detrimental factor to fracture healing.

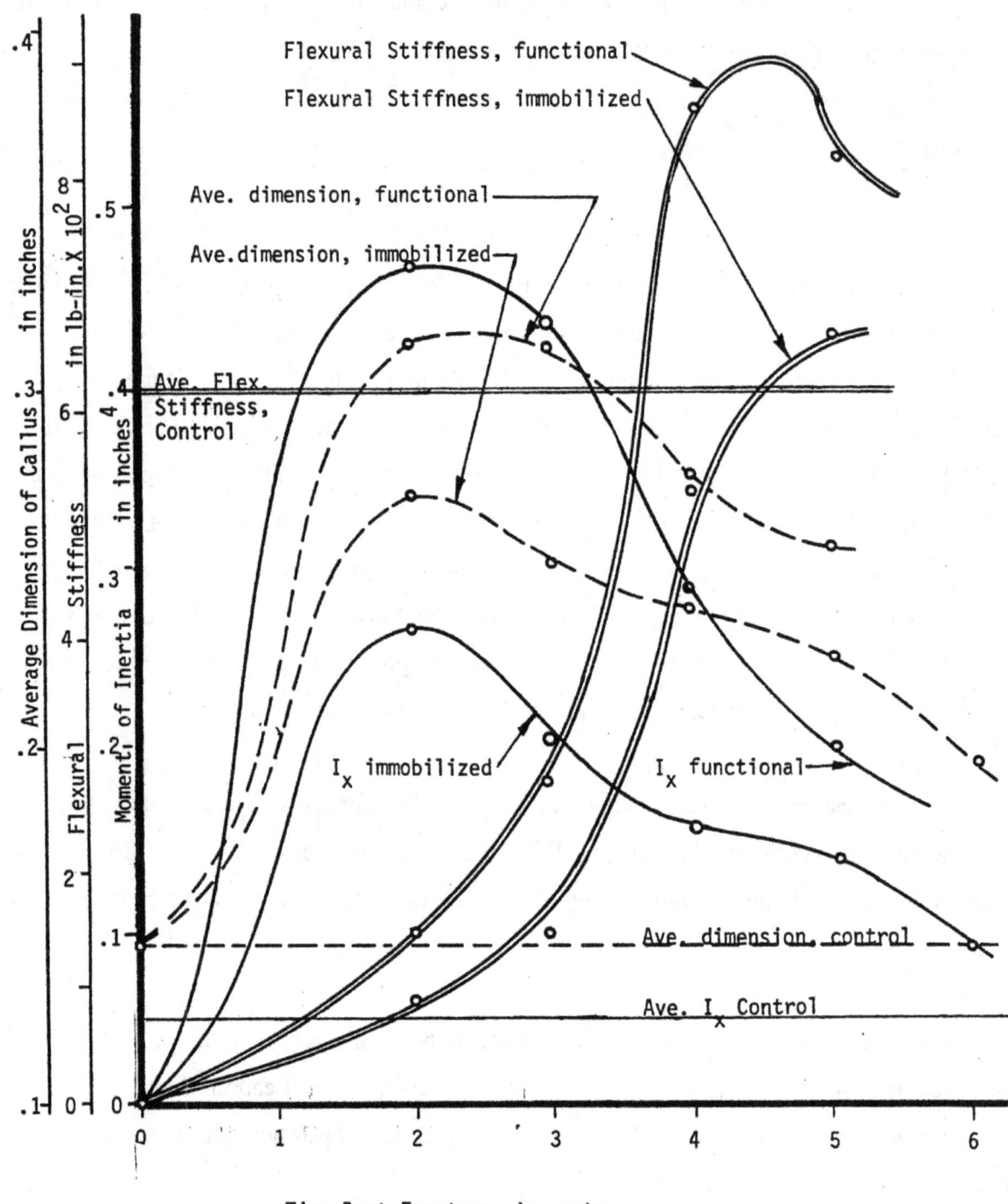

Time Post Fracture, in weeks

Figure 14 - Whole structure properties of healing fractures in rats.

Increased periosteal osteogenesis in the ambulatory group was considered an important factor in fracture healing. The reason for this increase is speculative. Perhaps through early function a greater inflammatory response provided a stronger stimulus to the osteoblastic layer of the periosteum thus stimulating new bone formation. [41] The change in appearance from the loosely woven bone of an early callus to a highly organized, compact and appositional structure in the functional group was primarily felt to represent the response of the fracture callus to the mechanical forces upon it. [72] The presence of many vascular connections between the appositional new bone and the original cortex which were not noted in the immobilized group was felt to indicate a pattern of increased circulation based on the active use of the involved extremity. [17,19,47,61-62]

The persistence of the fracture line on x-rays of specimens in the ambulatory group supports the idea that the rate of fracture healing cannot be accurately judged by x-ray studies alone. [39,67] These specimens were mechanically stronger than those in the immobilized group which had already lost their fracture line radiographically. Excess periosteal callus in delayed and non-union, appears on x-ray similar to the healing callus of functionally loaded limb, but the material properties are much different. [14,66,69,72]

The use of the pneumatic fracturing device caused little trauma to the soft tissues adjacent the fracture. This resulted in minimizing any de-vascularization that might occur to these tissues. Since the amount of trauma required to produce the fracture was kept uniform in both groups, differences in the rate and quality of fracture healing was not dependent upon this factor.

Local de-vascularization of the soft tissues has been shown to retard fracture healing and is an important consideration since the tissues adjacent the fracture callus provide the initial blood supply to it through vascular in-growth. [6,16,24,25,58,72] Likewise, the insertion of a loose, medullary pin did not in any way provide rigid immobilization or destroy the medullary circulation. [12,18,46] It did, however, allow for a uniform fracture gap and at the same time prevented any significant displacement and angulation.

Another consideration was the treatment of the specimen prior to the mechanical testing. Mechanical tests performed on specimens preserved in formalin or allowed to dry can give erroneous results .[36] This occurs through alterations of the organic components. .Drying of the callus will produce cracking along which a fracture can propagate. For this reason all specimens for testing were fresh and kept moist. Also removal of all soft tissue and periosteum about the fracture site will alter the load distribution during testing and effect results. This, however, was accepted since each specimen would be prepared in the same manner and only comparative differences would still be significant.

The three point bending test proved very satisfactory from the mechanical standpoint for this comparative study. The testing machine provided a rapid application of load and each fracture specimen could be tested as a structure without having to prepare a uniform machined sample. The diaphyseal portion of the rat femur was the most uniform, geometrically. Likewise, measuring deformation was made easier using a three point bending method than testing by axial loading. Weir, Bell and Chambers similarly tested the mechanical properties of whole rat femurs[70.] Our results were comparable consistent. with theirs and tended to validate our method of testing.

Because bone is a composite material exhibiting elastic, viscoelastic and plastic properties the results from any bending test should be open to interpretation. The formula $G=Mc/Ix$ calculates maximum stress in a homogenous, isotropic material within the elastic range. Bone and callus are viscoelastic, non-homogenous, anisotropic and the tests were carries out beyond the elastic range. The calculation of Ix for he callus assumes that the cross section was oval, thus the stress distribution was linear in the calculations. In reality, the stress distribution is non-linear (Fig. 15A) but the maximum stress reached in the plastic range is probably relatively proportional to the calculated ultimate stress values. As each fracture specimen is loaded at a constant rate the posterior part of the fracture callus comes under tension while the loaded side comes under compression. Under this condition, considerable plastic deformation occurs prior to failure. This material is termed ductile. Since failure is not sudden and uniform throughout the material, actual maximum stresses cannot be accurately calculated. But a calculates stress based on the maximum load carting capability of the bone in bending gives meaningful, effective stresses which are comparable between specimens.

The significance of the differences in material and whole structure properties of healing fractures subjected to functional loads compared to those without function far outweighed any errors in calculations.

The modulus of elasticity is not significantly different when comparing both groups at the same interval during fracture healing. The actual structural stiffness is significantly different however. This is consistent with the concept that the gross arrangement of the material at the fracture site is responsible for some of the differences n the mechanical properties of the structure tested rather than a difference in the composition of the material. The marked increase in the modulus of elasticity and decrease in ductility between the third and fourth weeks in both groups represented that phase of fracture healing in which rapid mineralization of the collagen present in the fracture callus occurs. This rapid increase in the stiffness of the fracture callus is of considerable importance in fracture healing. Without sufficient stiffness provided by either a stiff material in the callus or a large callus with high area moment of inertia, a healing fracture cannot withstand the constant loading occurring during weight-bearing ambulation. Simply, stiffness prevents angulation and rotation. Angulation and rotation during constant cyclic loading may result in a fatigue fracture or increase stresses due to the displacement causing failure which could usually be resisted by the specimen with only minimal angulatory displacement.

We believe that our conclusions are not in conflict with the literature, even though periosteal callus is generally thought to be a sign of poor healing. [23,28,37,67] Investigations studying the formation of periosteal callus noted delayed radiologic healing when compared to the direct bone formation in the endosteal callus formed following rigid fixation. [4,27,72] But the periosteal callus formed was formed without bony contact and was not mechanically tested. [4,72] In other investigations where the periosteal callus was allowed to develop without interference, healing was faster and stronger in the functional group without internal fixation. [43,67] In the description of its development, it has been noted that periosteal callus not only develops a large peripheral bulk but its best mechanically developed materials lie in the most peripheral areas of callus. The newest formed fibrocartilage develops near the center of the callus and pushes the older well-developed hyline cartilage outward along with the psseous "cap" on the most peripheral portion of the periosteal "hump". [72] Since the osseous tissues are continually being formed between the cartilage and osseous layers, the earliest remodeling of osseous tissues takes place in the most peripheral portion of the periosteal callus. Therefore, the materials in the callus best suited to resist stresses have been advantageously placed further from the neutral axis of the bone where the highest levels of stress in bending and torsion develop (Fig. 15A). Endosteal callus even of direct bone formation develops its best material in the early stages near the neutral axis. [39,40,47] After a fracture the ends of the bone fragments become necrotic. This necrosis increases with the use of internal fixation devices. [40] The necrotic ends of the fragments are the last portion to heal. [42,45] Thus, in the endosteal callus the material in the best mechanical position develops last (Fig. 15B). This probably explains the minor differences in calculated modulus of elasticity between the functionally treated and immobilized animals in our investigations. The material which is stressed the most characterizes the modulus of elasticity of the whole callus when calculated in the manner described. In the periosteal callus the best developed osseous tissue is located peripherally and thus contributed most to the calculation of material stiffness. In the case of endosteal callus, the centrally located and best developed material contributed least to the calculation of modulus of elasticity.

Summary

Various examinations and tests were carried out to study fracture healing under functional and immobilized conditions. .The effects of altering functional activities on the mechanical properties of a healing structure were studied. Fractures which were subjected to functional forces showed an increased ability to withstand significant loads prior to failure. Functional loading and unloading of the fracture along with the maintenance of normal muscle mass and vascular supply provided the most suitable environment for the development of strong union. The research of others [6,15,16,25,58, 72].

32

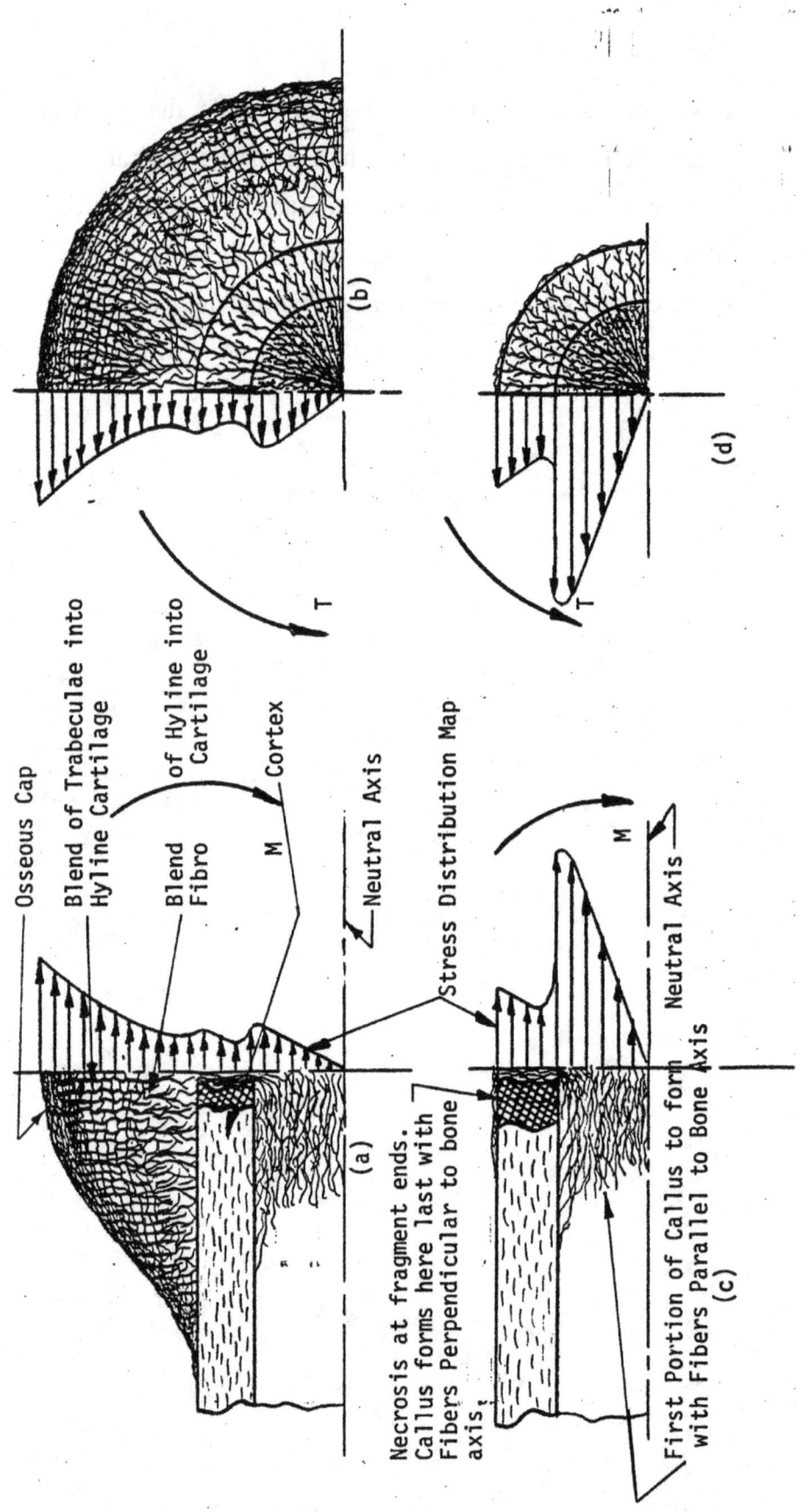

(b)

(d)

T

Osseous Cap

Blend of Trabeculae into
Hyline Cartilage

Blend
Fibro

of Hyline into
Cartilage

M   Cortex

Neutral Axis

Stress Distribution Map

(a)

Necrosis at fragment ends.
Callus forms here last with
Fibers Perpendicular to bone
axis¦

M

Neutral Axis

First Portion of Callus to form
with Fibers Parallel to Bone Axis
(c)

Fig 15. Hypothetical tensile stress distribution maps for periosteal (a) and endosteal (c) callus as a result of bending load M. as illustrated in longitudinal views. Axial views illustrate hypothetical shear stress distribution maps for periosteal (b) and endosteal (d) callus under axial torque load. The areas in all stress maps are equal, indicating the relative levels of stress required to resist equal moments. Note that the stress levels are much higher in the endosteal callus due to the reduced moment of inertia, which is not only a result of reduced callus mass, but also poor position of materials. The stiffest material in the periosteal callus is located most peripherally, thus it bears a disproportionate amount of stress. The result is the non linear distribution of the stress over the gradient of stiffness caused by variation in material properties within the cross-section (the softest materials toward the center). The whole callus is protected from high levels of stress until the center has developed the properties to withstand the loads of weight bearing by itself. Then the callus shrinks with remodeling.

Into the notion on bone changes associated with impaired circulation, disuse and denervation tend to support this concept. Although the fractures in both the immobilized and functional animals healed, those fracture in the immobilized group never reach pre-fracture mechanical strength within the period of time of this study. This indicates a decrease in the functional capability of the fracture callus in not solely dependent upon the integrity and vascular supply of the periosteum and soft tissue about it.; motion and local mechanical forces contribute significantly to the development of the callus. This is evident by the amount of cartilage produced. However, similar to the observations of others, [48] we could find no pertinent relationship between the development of the callus size and the size of the initial hematoma at the time of fracture. Differences in the callus were directly related to the differences in the rate and quality of fracture healing. Functional forces results in increased stability of the fractures in the ambulatory group. This was in part the result of remodeling. In bending and torsion the deforming stresses are greatest at the periphery of the callus. Resistance, therefore, to the tensile, compressive and shear stresses that develop must be high in the peripheral portion to prevent failure. The external callus with its large peripheral mass provides this stability. In short, the architectural arrangement of the substance of the callus, which is, closely dependent upon the mechanical forces across the fracture site can be more important than its material strength alone. For this reason the formation of a relatively large callus does not necessarily mean the fracture healing will be delayed or the quality of the healing process will be impaired; it simply provides an increase moment of inertia so that the functionality loaded bones are stiffer structures than the immobilized bones even though the material properties of each are similar.

The concepts of callus formation in immobilized and non-immobilized animals cannot necessarily be applied to those fractures treated by other methods. However, the importance of the function loads and adjacent muscle activity in prompting fracture healing in fractures treated by closed methods in stressed.

SECTION II

BIOMECHANICAL STUDIES ON FRACTURE STABILITY

Biomechanical Studies on Fracture Stability

The following biomechanical studies were conducted primarily for the purpose of obtaining information concerning the mechanical reasons that would explain the apparent success in the treatment of tibial shaft fractures with our below-the-knee functional brace. We sought explanations for the fact that the length and alignment of fracture fragments are maintained despite the subjection of the fractured limb to weight-bearing stresses shortly after the initial insult.

We contend that the elastic motions at the fracture site resulting from functional activities, does not interfere with osteogenesis. To the contrary, it appears to favorably enhance the reparative process. However, we recognize the importance of preventing major, permanent displacements between fragments with potentially adverse clinical and/or cosmetic consequences.

The laboratory studies offer explanations for the manner in which the functional braces provide this necessary stability.

EXPERIMENT 1:

The Load Distribution of the Functional Belowthe-Knee Fracture Brace

The original thought behind our use of the PTB design for treatment of tibial fractures was to unload the fracture site during ambulation through a mechanism similar to the patellar-tendon-bearing prosthesis 51. Thus the functional below-the-knee cast was designed 50. Once we developed the functional below-the-knee brace (Fig. [16A,B,C]), Page 38, it became evident that all the Toads externally applied to the limb could not completely by-pass the fracture site simply because at push off the ground reaction Toads are not transferred through the brace." The method still appeared to work well clinically, but our rationale became questionable. Attention was paid to the comments of patients who rarely felt pressure in the area of the PTB molding or tibial condyles. Many patients noted pressure on the calf where the bulky, soft tissue mass of the gastroc-soleus muscle was tightly molded by the brace. It became evident that answers to the questions of the load bearing function of the brace and the distribution of loads could not be adequately answered through clinical observations. Therefore, the following laboratory studies were performed in an attempt to better understand the true function of the functional below-the-knee brace.

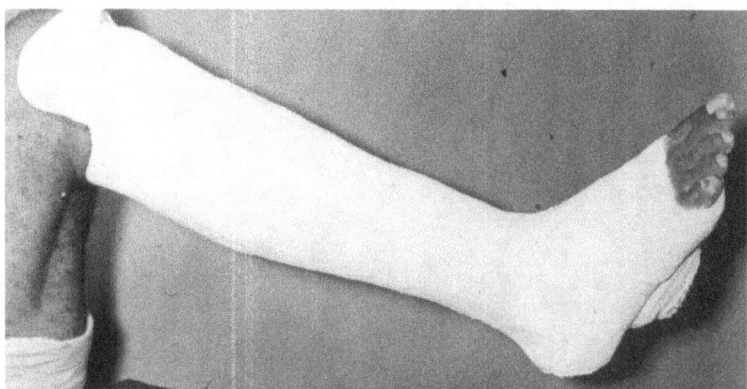

Fig. 16A.

Below the knee functional cast which allows freedom of motion of the knee joint and weight bearing ambulation

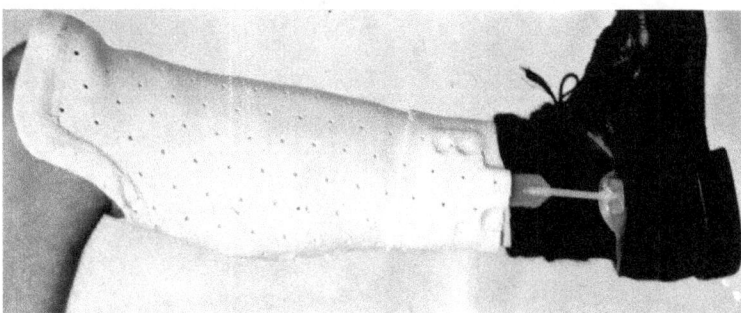

Fig. 16B.

Functional below the knee Orthoplast brace which permits freedom of motion of all joints of the injured extremity.

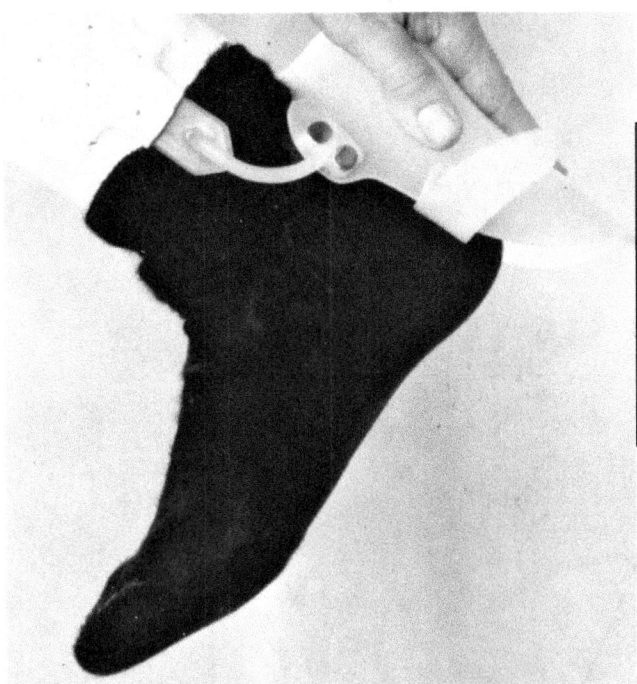

Fig. 16 C.

Flexible ankle insert which facilitates donning and doffing of the shoe.

Methods and Procedures

Efforts to utilize transducers to measure the pressures developed between the brace and the underlying skin failed. The variations in the compliance of the underlying tissues and the dynamic changes during, muscle contractions precluded calibration., Therefore, we divided the brace into proximal, mid and distal sections which were held together by rigid Orthoplast struts. Strain gauges were applied" to the struts to measure the loads transferred between the sections. Consequently the load bearing function for each section of the brace could be individually studied. For the purpose of calibration and production of a theoretical model to analyze the results of the in vivo testing, a conical model of three graduated, truncated cones of wood connected by load transducers was constructed with a surrounding Orthoplast sleeve similar to the Orthoplast brace. From these measurements the total loads could be calculated and compared to the total loads applied to the system (as measured by strain gauges in the ankle joint of the brace). The mechanical properties of the Orthoplast were determined by cantilever bending tests with strain gauges applied.

Results and Conclusions

The results of tests on the Orthoplast material indicated that the ultimate tensile strength averaged 5,000 psi; the elastic modulus at 300%; elongation was 2,300 psi; the distortion temperature was 136° Fahrenheit; the melting temperature 221° Fahrenheit and the elongation averaged 475%.

The conical model instrumented with strain gauges gave the results shown in Fig. 17. The fully instrumented brace was then applied to a subject without a fracture and calibrated. The subject then ambulated on a treadmill while readings were taken (Table IV).

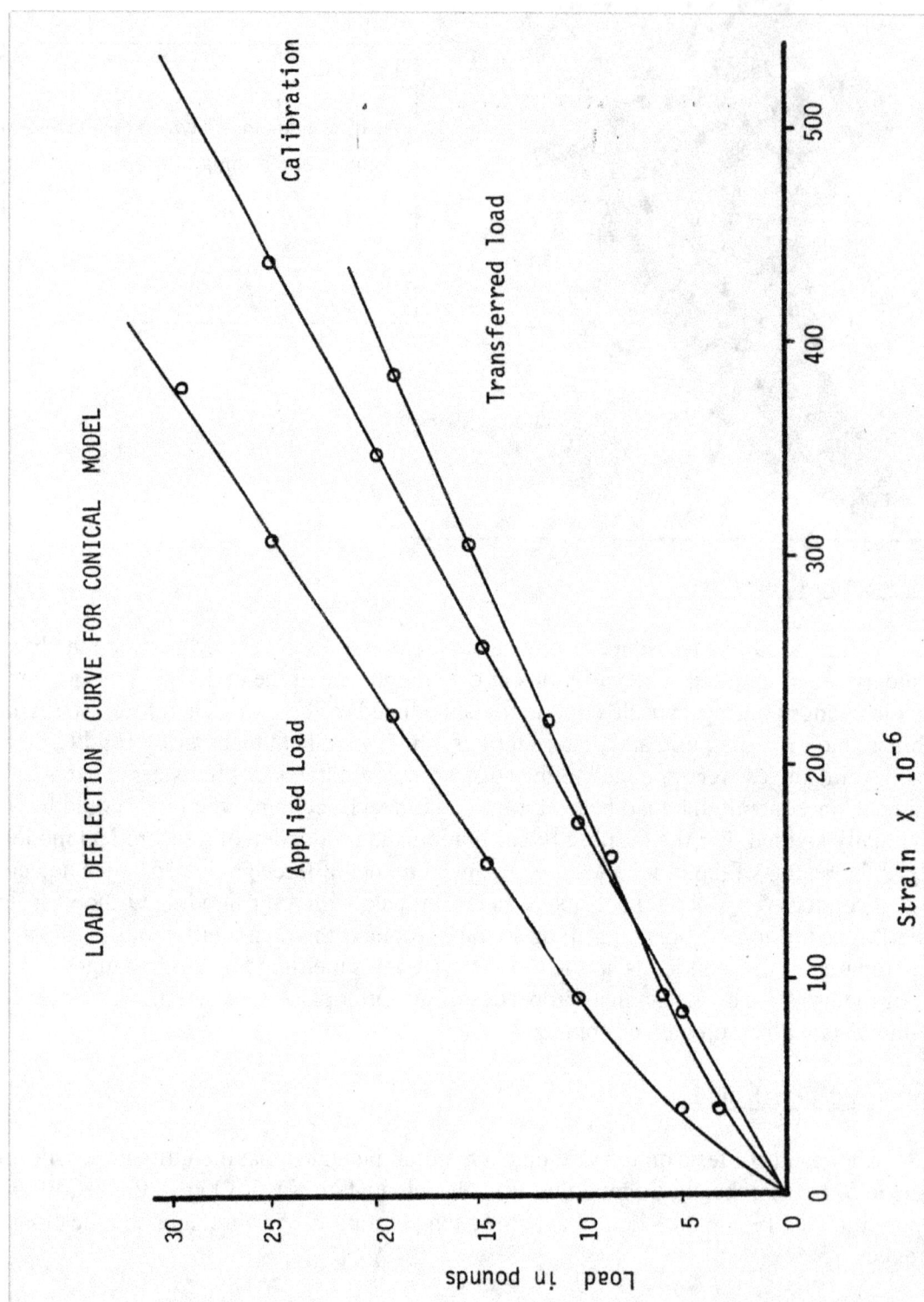

Once the system was refined to give accurate results, a patient with a tibial fracture which allowed significant pistoning of the fragments was fitted with the brace. His ambulation on the treadmill (Fig. 18) gave the following results (Fig. 19). The forces on each section of the brace were calculated and are tabulated in. Table V. A comparison of stresses taken by the brace compared to the stresses which were calculated by the theoretical model are displayed in Table VI. The final percentages of external loads transferred through the brace are tabulated in Table VI ,It should be noted that the results are a measurement only of the axial load carried by the brace. No measurements of circumferential stresses were attempted by these techniques.

Results of tests on normal subjects indicate that approximately 17% of the load normally carried by the leg is transmitted through the brace and that roughly 85% of the transfer occurs in. its proximal half. Tests on a patient with a nonunion ambulating in the instrumented functional below-the-knee fracture brace indicated that the maximum load which could be expected to be borne by the brace was approximately 26% of the load borne by the leg. However, under these conditions the patient was "jacked up" in the brace so severely that he experienced pain upon ambulation. Under the normal adjustment of the fracture brace, the patient transferred a maximum of 20% of the load borne by the limb through the brace. This eliminates the possibility of the brace functioning to significantly offload the bone. The results further indicated that only small bending moments are taken by the brace.

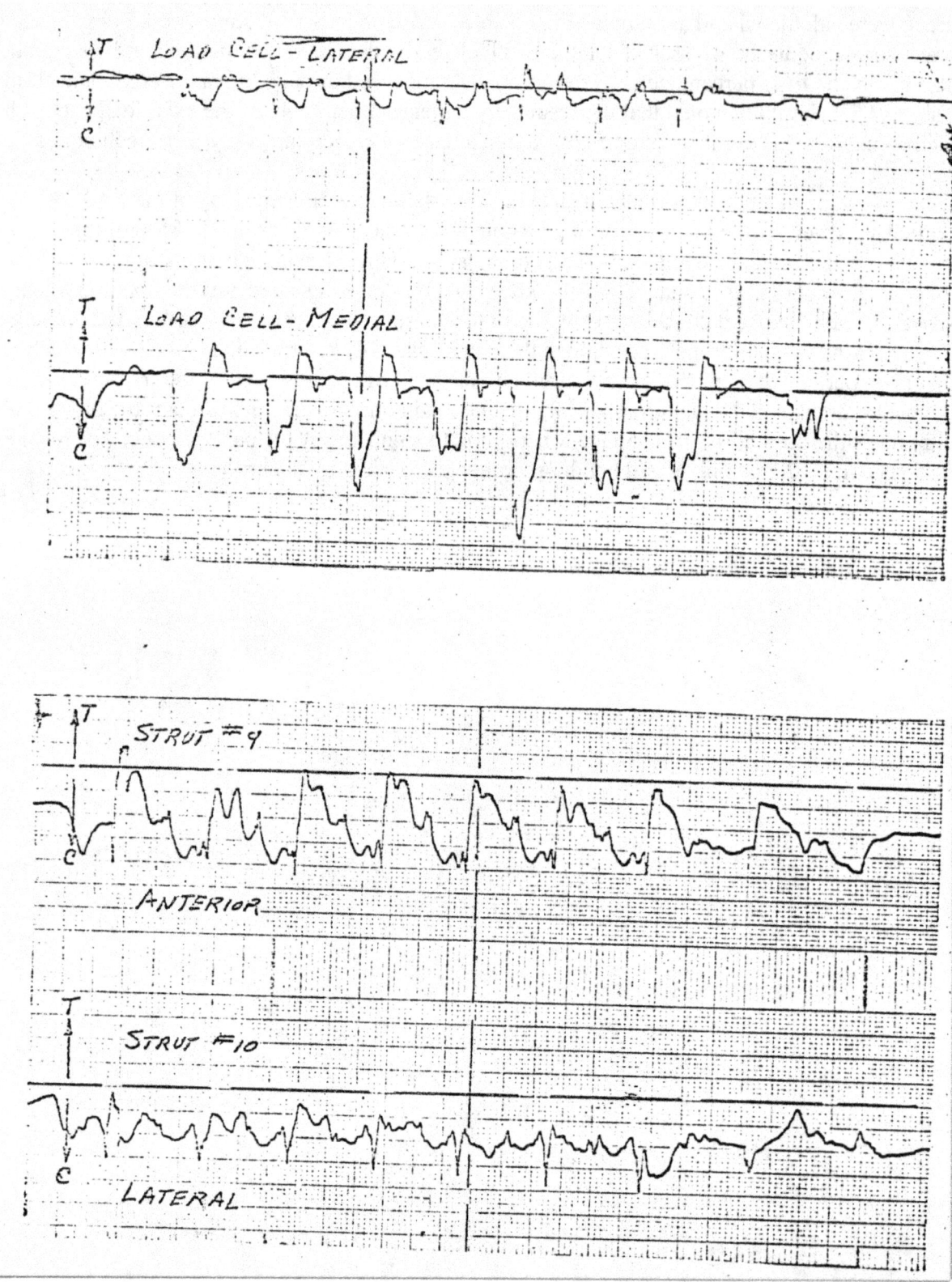

Fig. 18 – Example of strain gage outputs on the chart recorder for two of the twelve struts on the instrumented fracture brace. The load cell readings are of the total external loads applied to the brace which were transmitted thru the ankle joints of the brace at the patient ambulated on a tread mill.

TABLE   IV

STRUT FORCES AT HEEL STRIKE

Strut Location	Top Section	Middle Section	Bottom Section
Anterior	.118*c	.238c	.234c
Lateral	.74 c	1.34 c	1.41 c
Posterior	1.21 c	1.36 c	------
Medial	1.57 c	1.23 c	1.80 c

* Loads in kg.
c- compression, t- tension

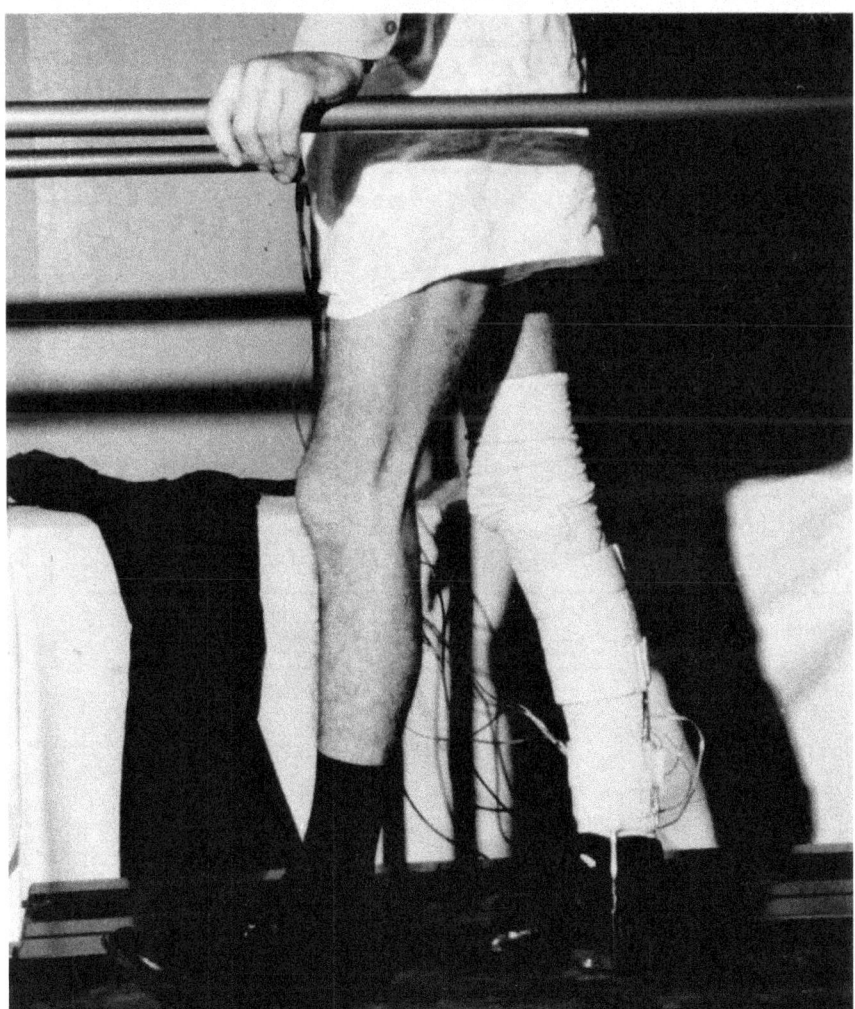

Figure 19 – Patient with unstable non-union ambulating on the treadmill, wearing the instrumented fracture brace.

TABLE V

FORCE TRANSFERRED THRU EACH LAYER

	Top	Middle	Bottom
Total Force Transferred Thru Brace (kg)	10.5	11.8	13.2
Total Force Transferred Thru Each Layer (kg)	5.1	10.2	11.8
% of Total Force in Each Layer	48%	86.5%	89.5%

TABLE VI

EXPERIMENTAL AND CALCULATED VALUES

	% of Load Borne by Each Layer		
	Measured on Patient	Experimental Model	Calculated Theoretical
Top	48%	40%	42%
Middle	39%	39%	41%
Bottom	3%	27%	29%

TABLE VII

LOAD BEARING FUNCTION OF BRACE ON PATIENT WITH NON-UNION

Load Application	Total Load Transferred Thru Brace (lbs)	% of External Load on Braced Leg, Taken by the Brace
Full Weight on Braced Leg	23.5	21%
Equal Weight on Each Leg	15.5	27%
Ave. Heel Strike With Adjusting Screws Too Tight	39.0	26%
Ave. Heel Strike With Adjusting Screws Normal	29.5	20%

The functional below-the-knee brace has proven itself clinically as a viable alternative in the treatment of tibial fractures. It has some mechanical effects which alter the stresses at the fracture site. It is this effect that must ultimately be delineated. However, it seems obvious that other factors must be involved in the effective stabilization of tibial fractures.

EXPERIMENT 2:

Hydraulic Effects of Soft Tissues

Since the previous study indicated that the functional below-the-knee brace unloads the limb by approximately 17%, explanations have been sought for the apparent stability experienced by tibial fractures treated by this method. we have advanced the concept that an "incompressible fluid" or hydraulic function of the water-like soft tissues encapsulated in the fracture brace provides an important stabilizing effect 54 (Fig. 20) Page 45. However, the hydraulic effect of soft tissues can only prevent shortening, not angulation, under this concept. The concept of the hydraulic effects of soft tissues is not new. Other investigators have looked at dynamic and static tissue pressure effects of a similar nature in other portions of the body 214. The following studies attempt to experimentally demonstrate the success or failure of this effect in a practical manner.

Procedure

A fresh above-knee amputation specimen was stripped of all soft tissues within a space of about 4 inches long at approximately the mid-shaft of the tibia and fibula. A fracture of the tibia was then created by a saw cut at an oblique angle. A segment of the fibula was also removed to allow movement of the tibial fragments. A transparent, acrylic brace was applied on the specimen so that the fracture fragments could be observed. The void created by the removal of the soft tissues was filled with a clear, homogeneous gelatin so that the material filling the tissue gap could only prevent shortening by its incompressible fluid effect. The leg was then placed in a fixture and loaded (Fig. 21) Page 47. The fixture allowed no loads to be directly borne by the fracture brace. The load was applied to the proximal fragment of the tibia and borne distally by the tibia via Steinman pins. The brace touched only the soft tissues and gelatin. Slots in the brace allowed the brace to slide up or down without applying loads to the distal pins in the shaft of the tibia.

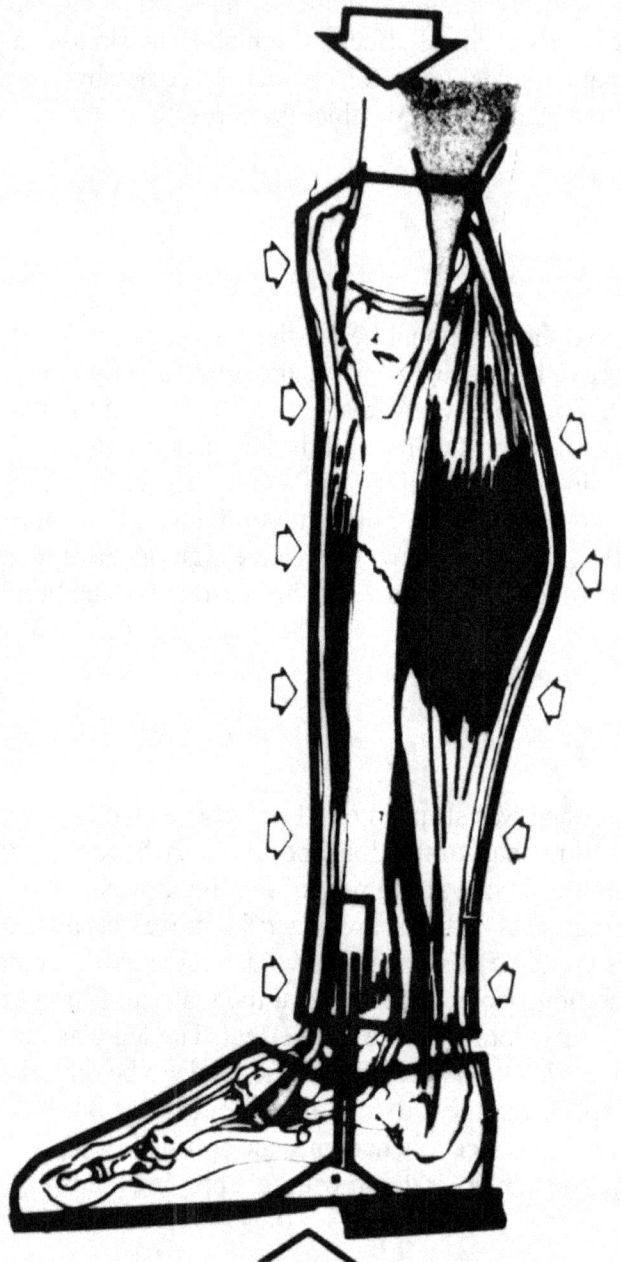

Figure 20

The advantageous anatomy of the leg for hydraulic, incompressible fluid effect of the soft tissues when load is placed on a fracture leg fit with functional below the knee cast or brace.

The first specimen was loaded with the same kinetic energy as would be absorbed at heel strike by a 150 pound subject walking normally. The loading technique, however, permitted the instantaneous forces to peak too high and the pressures in the gelatin-system reached much higher peaks than anticipated. ~In the absence of a skin covering the defect, the experimental brace developed leaks at high pressures and the gel extruded out allowing the leg to shorten one and one-sixteenth inch. This indicated the importance of the hydraulic mechanism in the prevention of shortening in this experimental specimen (Fig. 22) Page 58. A second specimen was loaded with a damped harmonic oscillating load in such a manner that the peak force reached by the system was equal to 150 pounds. The transparent brace was sealed.

47

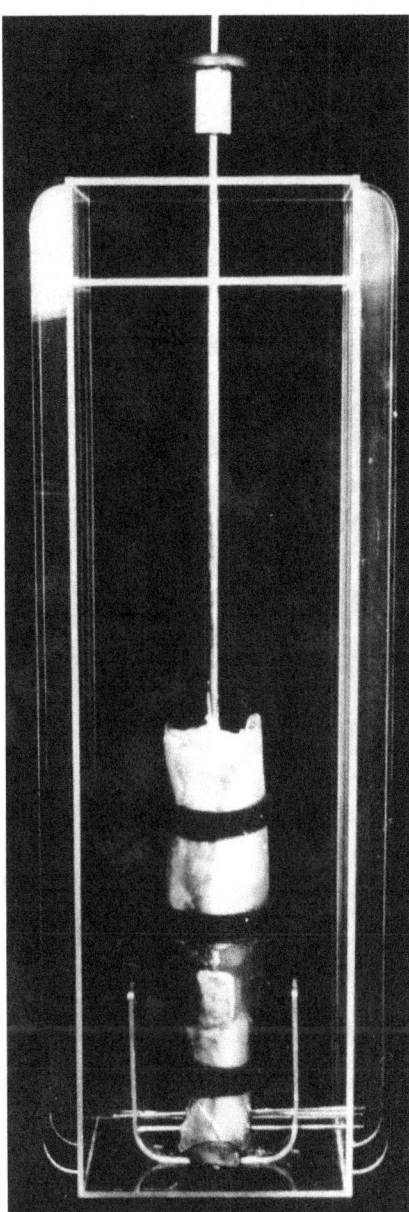

Fig 21A – Fresh A/K amputation in clear fracture brace with tibia and fibula fractured mounted in fixture for load applications.

Fig 21B – Close up of the leg showing section of the gelatin replacing soft tissues

Results and Conclusions

The system held intact and shortened 1/16 inch under static load of 25 pounds and 1/16 inch more with the first cycle of load. Repeated loading cycles produced no additional, permanent shortening We suspected that the shortening observed was due to the loose fit of the brace to the leg. Since the acrylic brace cannot be molded directly to the leg the fabrication had a much less "intimate" fit than a properly applied Orthoplast brace.

This experimental study demonstrated that the anatomy of the leg is advantageous for encapsulating the soft tissues within a functional below-the-knee fracture brace. Large and bulky muscles become tendinous before crossing the knee and ankle joints and are tightly held against the underlying bones by crural ligaments and fasciae. The tissues cannot extrude across the knee or ankle joint to allow displacement of soft tissues. Thus, shortening of the limb in the presence of a loaded fracture under these conditions cannot occur. These laboratory findings suggest that the hydraulic effect of soft tissues in a fracture brace could be an important mechanical contribution to the prevention of shortening.

EXPERIMENT 3:

Brace Versus Soft Tissues

The previously described experiment concerning load bearing capability of the functional below-the-knee fracture brace indicated that the soft tissues must play an important role in bearing the axial and bending loads placed on the limb during normal ambulation. The second experiment demonstrated that the hydraulic effect of the soft tissues was likely to be effective in the prevention of shortening. However, hydraulics can only explain prevention of shortening and bearing of axial loads applied to the limb. Angulatory deformities must be prevented by other means.

The following laboratory investigation attempts to compare the stability of a fractured tibia with and without a brace. In order to demonstrate the approximate relative importance of the hydraulic effect of the soft tissues to the inherent stability of the soft tissues these tests were repeated with damage selectively caused to various soft tissue structures. This was done to isolate their specific roles in stabilizing tibial fractures. One such structure is the interosseous membrane.

Procedure

A fresh above—knee amputation specimen was fractured by drilling holes in the tibia and completed by a blunt blow with a hammer; a similar fracture was created in the fibula. Nails were placed in the proximal and distal fragments of the tibia to serve as reference points for evaluation of angulation and length measurements. A fracture brace was applied. The leg and brace were held in a fixture so that the foot and knee could not appreciably move (Fig. 23). Static moments were applied to the horizontal and vertical (medial-lateral) planes as well as static loads. Forces were applied first to the leg without a fracture brace to measure the "fit" of the fixture and the ligamentous laxity of the limb. Then the loads were applied to a limb in a fracture brace with the artificially produced fracture. The loads were re-applied to the limb without the fracture brace (Fig. 24) .Finally, the interosseous membrane was stripped from the tibia with a fascial knife, damaging as little of the other soft tissues as possible. Loads were re-applied and deflections compared. The limb was dissected upon completion of the experiment to note the damage to the interosseous membrane and other soft tissues.

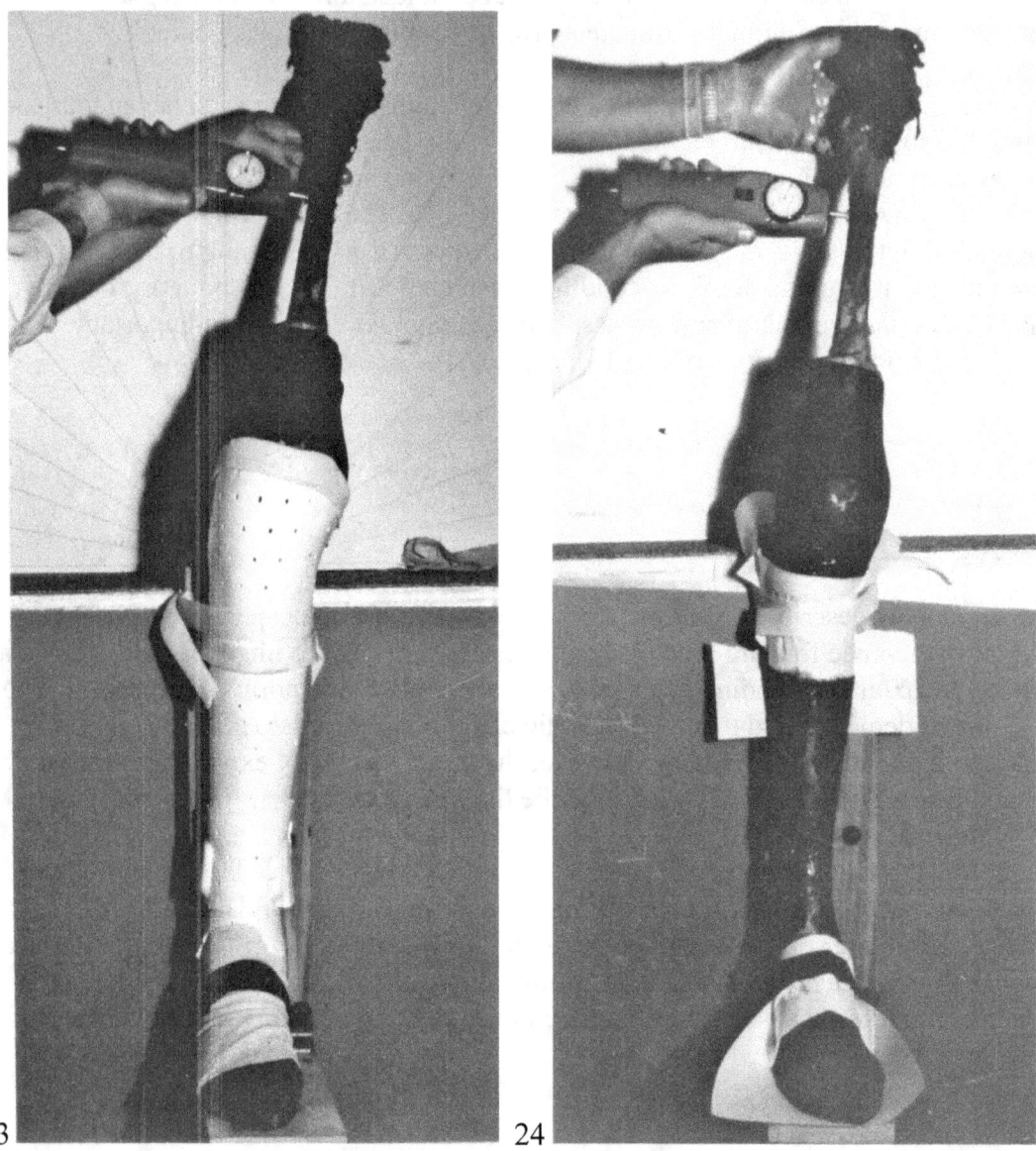

Figure 23-   Fresh A/K amputation in fracture brace with varus moment applied through femur.

Figure 24-   A varus moment applied to leg without the fracture brace.

Results and Conclusions

X-rays and visual observations indicated that the limb was much more stable with the brace than without it, particularly in regard to angulation (Fig. 25). Rotatory deformities were the least stabilized by the brace and recovery was more complete and rapid with the brace than without it. In the first case, loading of the limb with the interosseous membrane intact and then again with a damaged membrane showed little change in deflection (Fig. 26). Our preliminary impression was that the membrane was of relatively little importance compared to that of the brace. But when the limb was later dissected it was discovered that some of the membrane was intact at the fracture site holding the comminuted fragments together (Fig. 27).

A second limb demonstrated similar responses to those first observed before the interosseous membrane was stripped. But following the stripping of the membrane the limb lost stability (without a brace) as indicated by the x-rays and data chart (Fig. 25). Documentation of the successful stripping of the membrane in the second experiment was done by dissection (Fig. 28). A third limb was used to compare the stability of the fracture in the brace .with and without the integrity of the interosseous membrane. The x-rays indicated negligible changes in shortening at the fracture with or without the interosseous membrane in the brace stabilized extremity (Fig. 29). However, once the brace was removed the limb shortened. Assessment of membrane damage was made again by dissection.

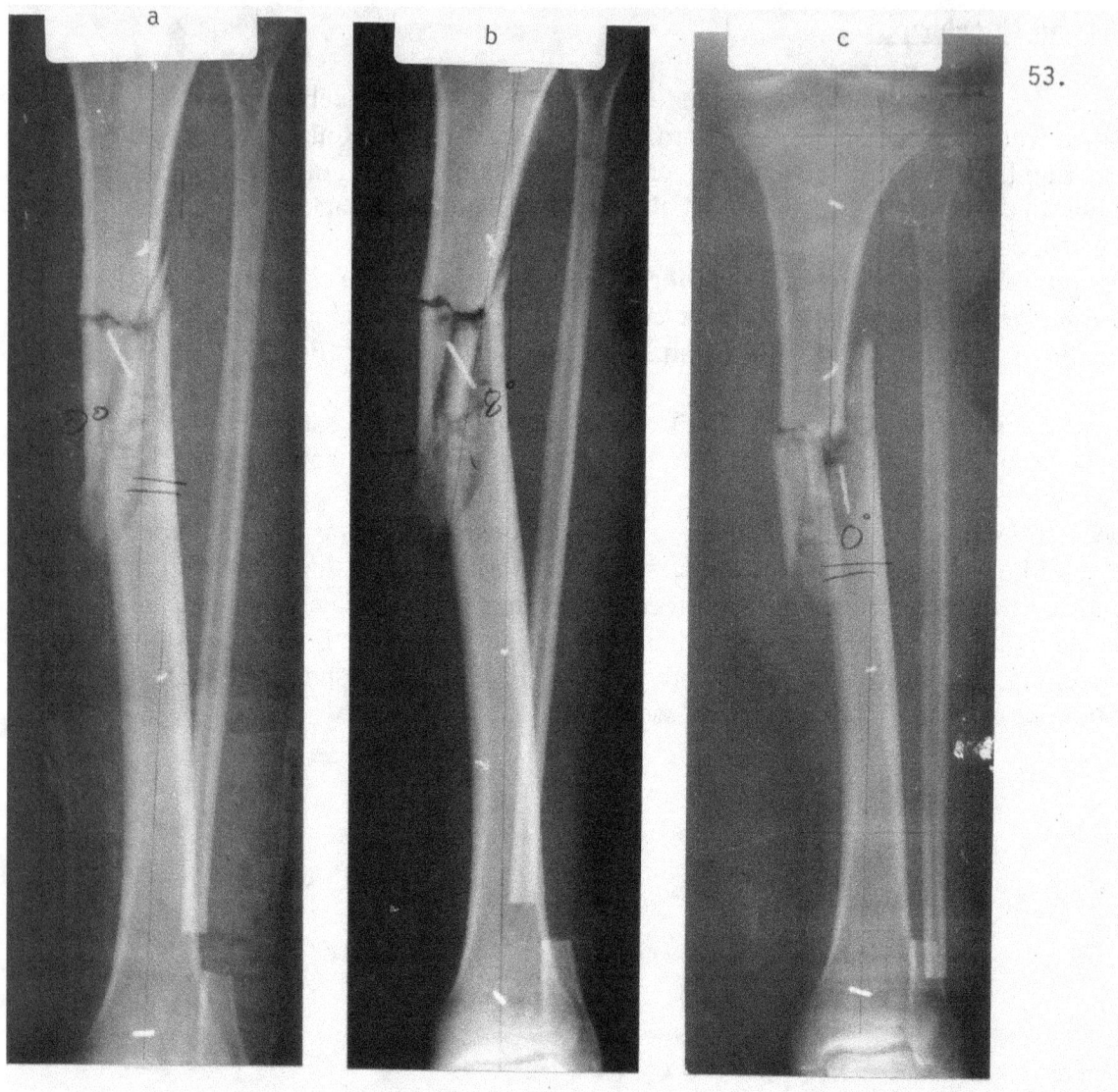

53.

CONDITION OF LIMB	AXIAL LOADS					VALGUS MOMENT, 5ft-# ANGULATION		AXIAL MOMENT, 3 ft-# ROTATORY DISPLMT.
	LOAD lbs.	SHORTENING abs., mm	SHORTENING relative, mm	REFERENCE	ANGULATION from X-ray,°	External,°	X-ray,°	External,°
Before Fx	0	0	———	———	0	———	———	———
Initial Injury w/o brace	0	10.68	———	———	.5	———	———	———
Reduced in brace	0	.51	10.15	Amount of reduction	7.0	———	———	———
Loaded in brace	60	2.03	1.52	Due to load	3.0	5	8.5	40
Loaded w/o brace	50	6.10	4.07	Greater than w/ brace	11.0	15	12.5	85
Loaded w/o membrane	45	35.55	29.45	Greater than w/ membrane	*	20	15.0	85

*Not applicable because a counter force was required at fracture site when loaded to prevent complete colapse of the limb.

Figure 25 – X-Rays of leg in Fig. 23 with Axial load applied in fracture brace (a), without fracture brace (b) and with interosseous membrane stripped from tibia (c). without fracture brace (e) and after interosseous membrane was stripped (f).

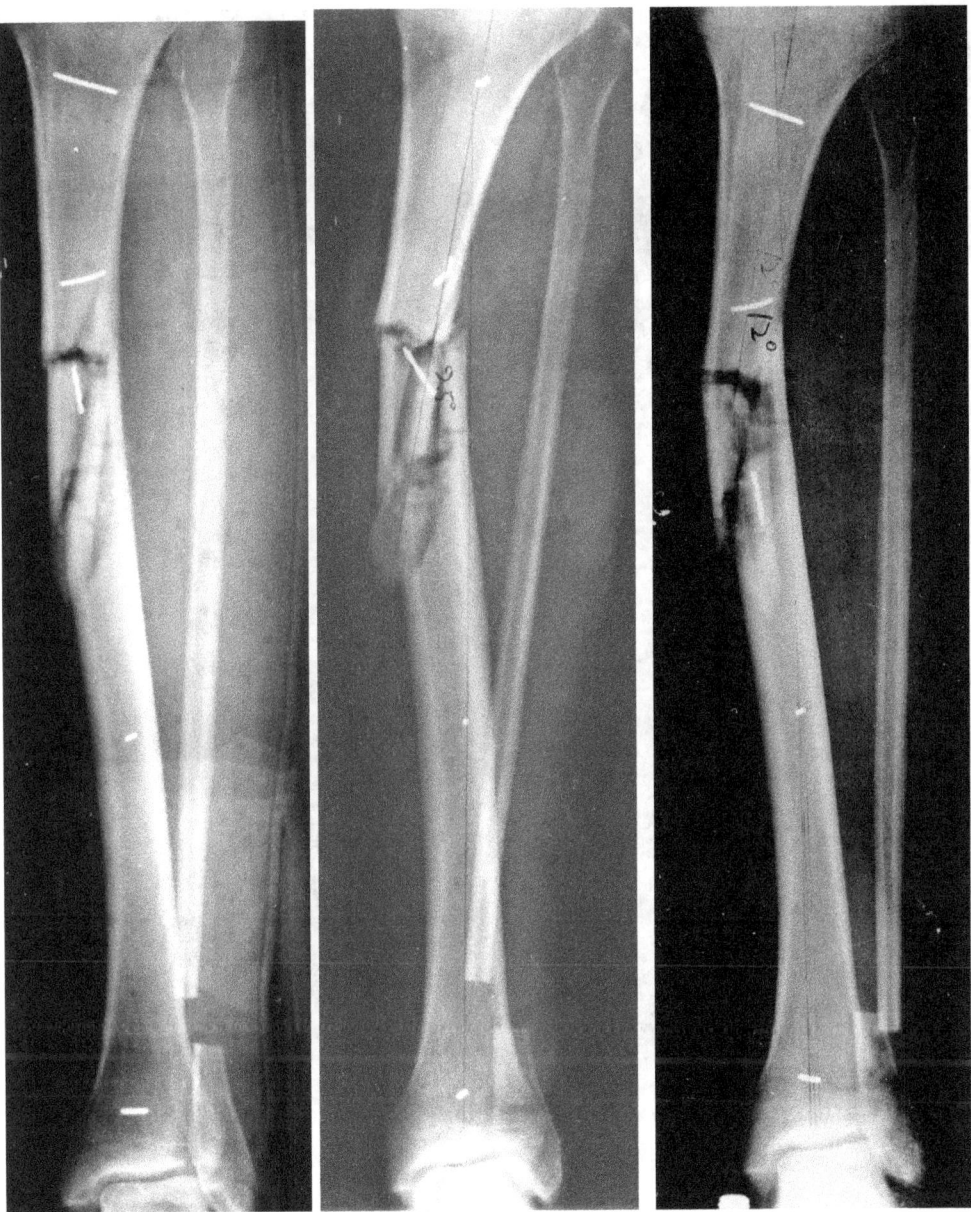

Figure 25 – With axial torsion, displacement is seen by changes in the projected lengths of the nails in the proximal fragment in the fracture brace (g), without the fracture brace (h) and with the membrane stripped (i). The measured displacement and loads are in the table. Absolute shortening is measured differences from nail locations before fracture was created. Relative shortening measures relative differences between arbitrary references as noted.

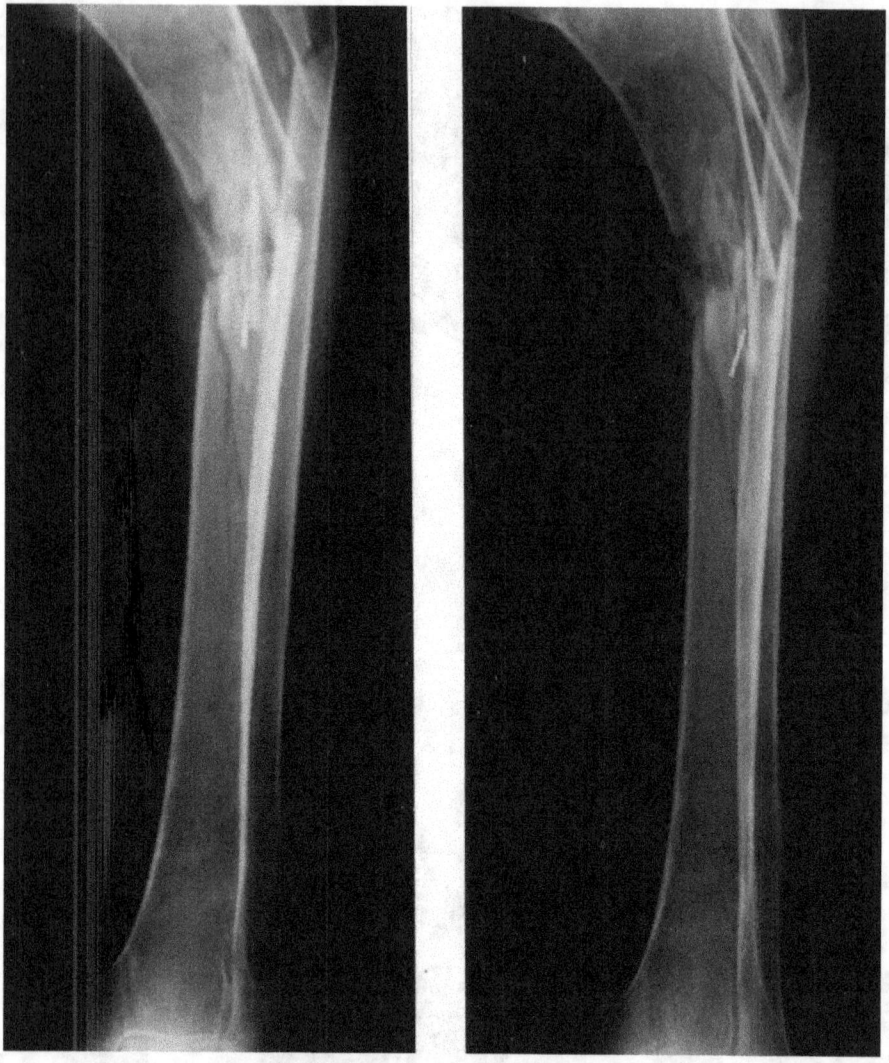

Figure 26 – Axial load applied before (a) and after (b) interosseous was damaged

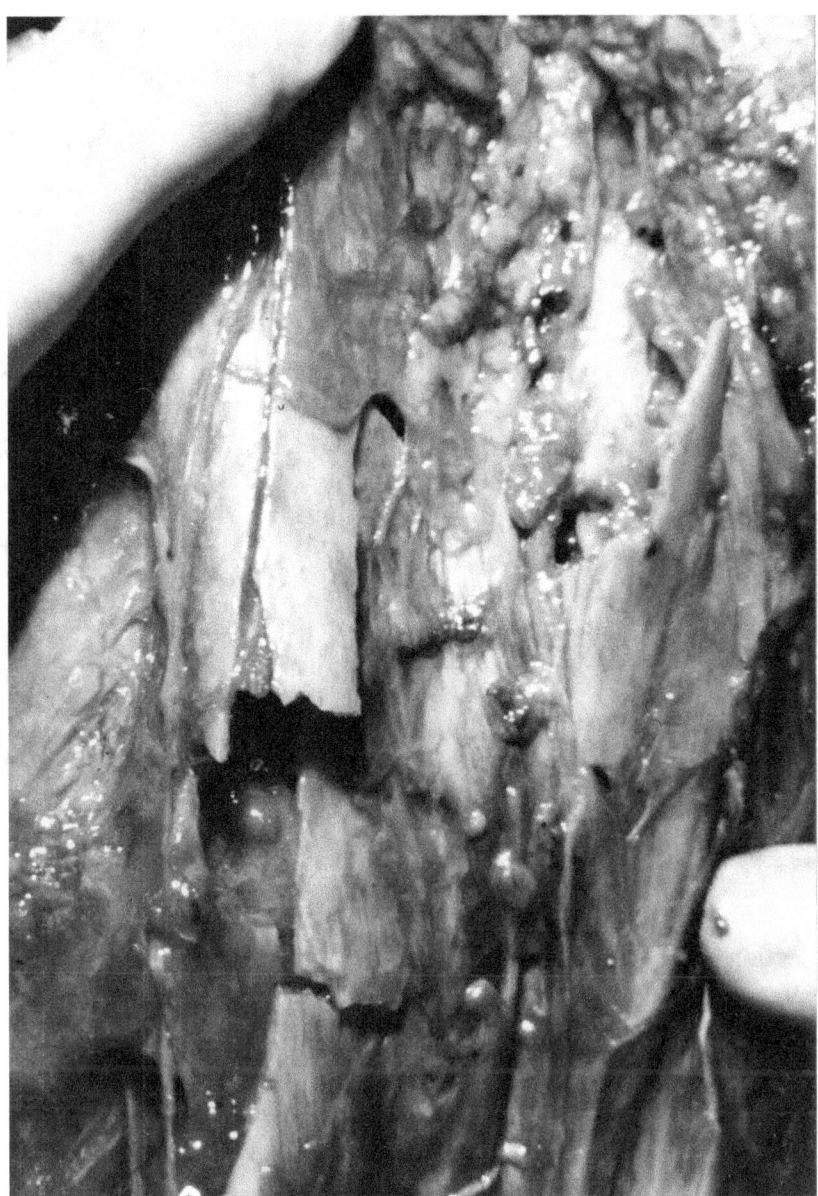

Figure 27 – Interosseous membrane intact at fracture site.

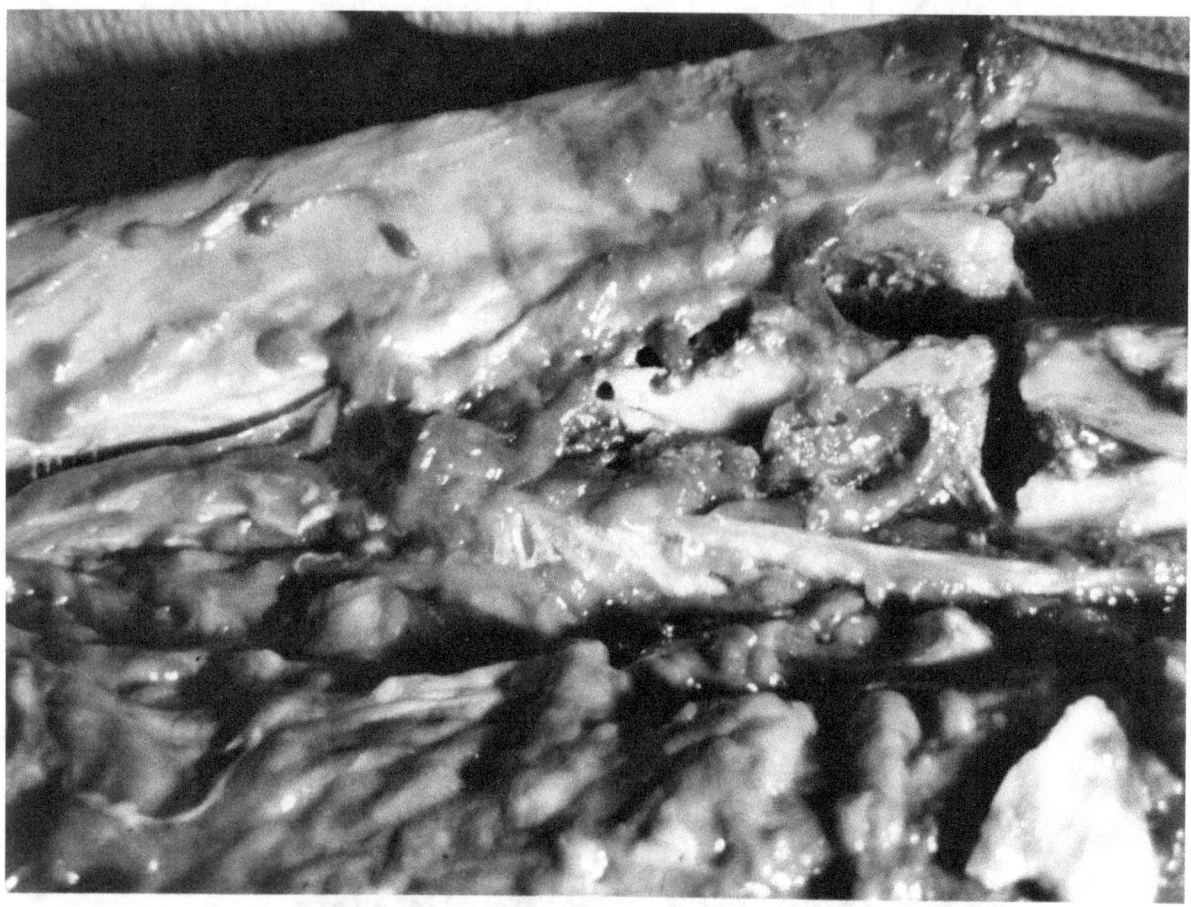

Figure 28 – Dissection at fracture site of limb showed fairly complete stripping on interosseous membrane from the tibia

Note: On the second limb, null measurements were made in the fixtures without a brace and without a fracture under identical loading conditions to those which were used in the subsequent tests. However, the limb in the brace with an artificially created fracture deflected less under a varus moment than did the limb without a fracture or a brace indicating possible control of ligamentous laxity by the brace. Thus, the null readings were considered to be meaningless and only relative values were measured from the radiographs. Once the fracture had been reduced in the fracture brace, even under loads, the shortening never reached the initial shortening observed before the brace and loads were applied.

It appears, therefore, that the below-the-knee functional brace is a most important stabilizing factor. Shortening and angulation were adequately prevented by its use. The interosseous membrane also seems to play an important role in determining initial and subsequent shortening and angulation.

Speculations in Reference to Clinical Application

Through an effective hydraulic mechanism, shortening of a fracture in a brace could be prevented as long as the fit of the brace is maintained until intrinsic stability develops from the healing of soft tissues and bone. However, if the fit of the brace is lost due to reduction of edema present at the time of its application and atrophy of the surrounding musculature before intrinsic Stability develops, shortening could probably recur during weight-bearing ambulation. However; if the interosseous membrane is intact, no additional shortening could take place as suggested by the laboratory studies. These extrapolations could explain our observations that tibial shaft fractures treated with a functional brace do not experience shortening beyond that present at the time of the original injury 1

Discussion

The functional below the knee brace seems to provide adequate stabilization against angulatory deformities. However, studies indicated that the brace bears little bending loads. Therefore, it must be assumed that the soft tissues, in conjunction with the brace, provide bending stability. The interosseous membrane has also demonstrated great relative importance compared to other soft tissues in stabilizing a fracture under laboratory conditions. So, it seemed logical to initiate studies of the interosseous membrane to better understand how it might contribute to the stability of tibial fractures.

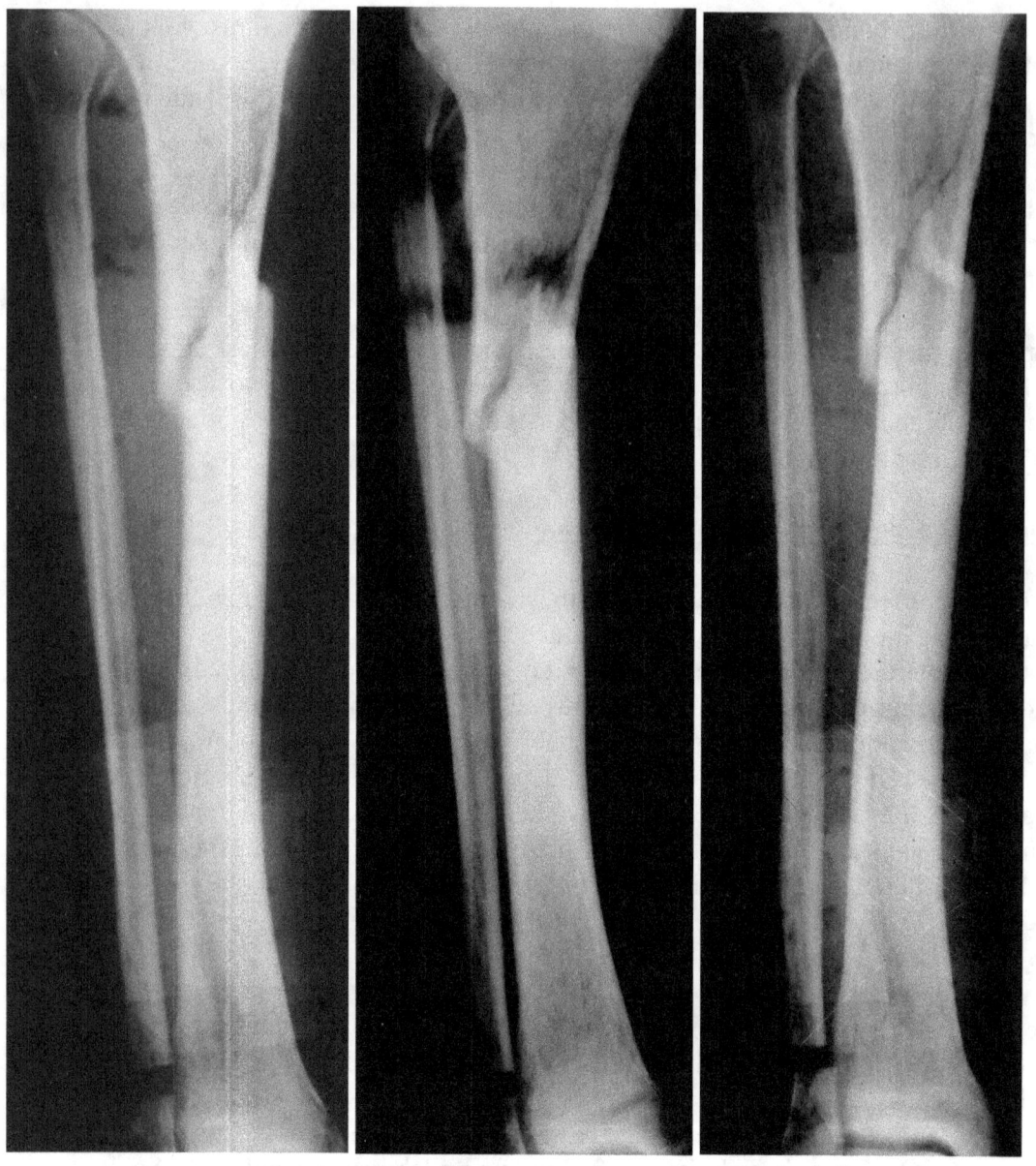

Fig 29A.   Fig 29 B.   Fig 29C.

Figure 29 A-B-C. Under axial load with the interosseous membrane intact, in the brace (A) and out of the brace (B). Note Typical angulation of tibia the fibula when the interosseous membrane "hinge" in intact After interosseous membrane was stripped, (C) (Continued)

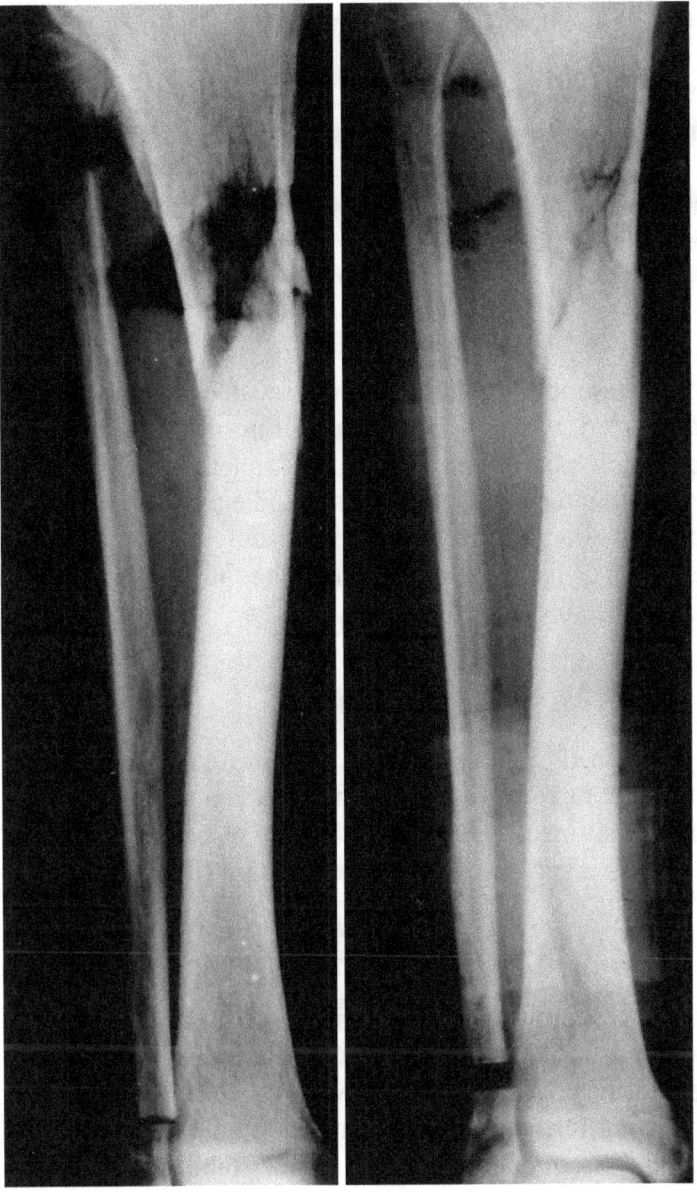

Fig 29D.   Fig 29E.

And out of the brace (D).

Without the interosseous membrane intact, the tibia angulates away from the fibula.

However when he brace is re-applied (e) the length is maintained under load for numerous re-applications of the brace.

EXPERIMENT 4:

The Role of the Interosseous Membrane

The previous experiments have indicated that the interosseous membrane could be an important soft tissue structure in stabilizing tibial fractures.

Background

The interosseous membrane has been described functionally as a muscle attachment site which probably bears none of the external loads of weight—bearing under normal conditions [20,21,31]. In the presence of a fracture, it has been hypothesized that it might function as a hinge or as a stabilizing tension member [7]. We found the interosseoue membrane to be a highly unidirectional, fibrous composite structure. Its fibers run from the tibia distally to the fibula at approximately 20° to the long axis of the tibia. Its thickness was found to be fairly uniform at 75 microns and consistent in numerous fresh specimens (Fig. 30).

Procedure

Formalized specimens of legs were stripped of all soft tissues with the exception of the interosseous membrane and the periosteum. Fractures were created in the tibia and fibula of each specimen under controlled laboratory conditions. .The specimens were placed in a fixture which allowed no varus or valgus deformities and subjected to vertical loading (Fig. 31).

Results and Conclusions

The amount of shortening at the fracture site was minimal when the interosseous membrane and the periosteum were intact regardless of the location of the fractures in the tibia and fibula. The relative location of the fractures and their displacement does, however, determine the amount of membrane, which is actively resisting further displacement. The relative location of fractures influences the amount of initial and subsequent angulation and the stabilizing effects of the interosseous membrane (Fig. 32).

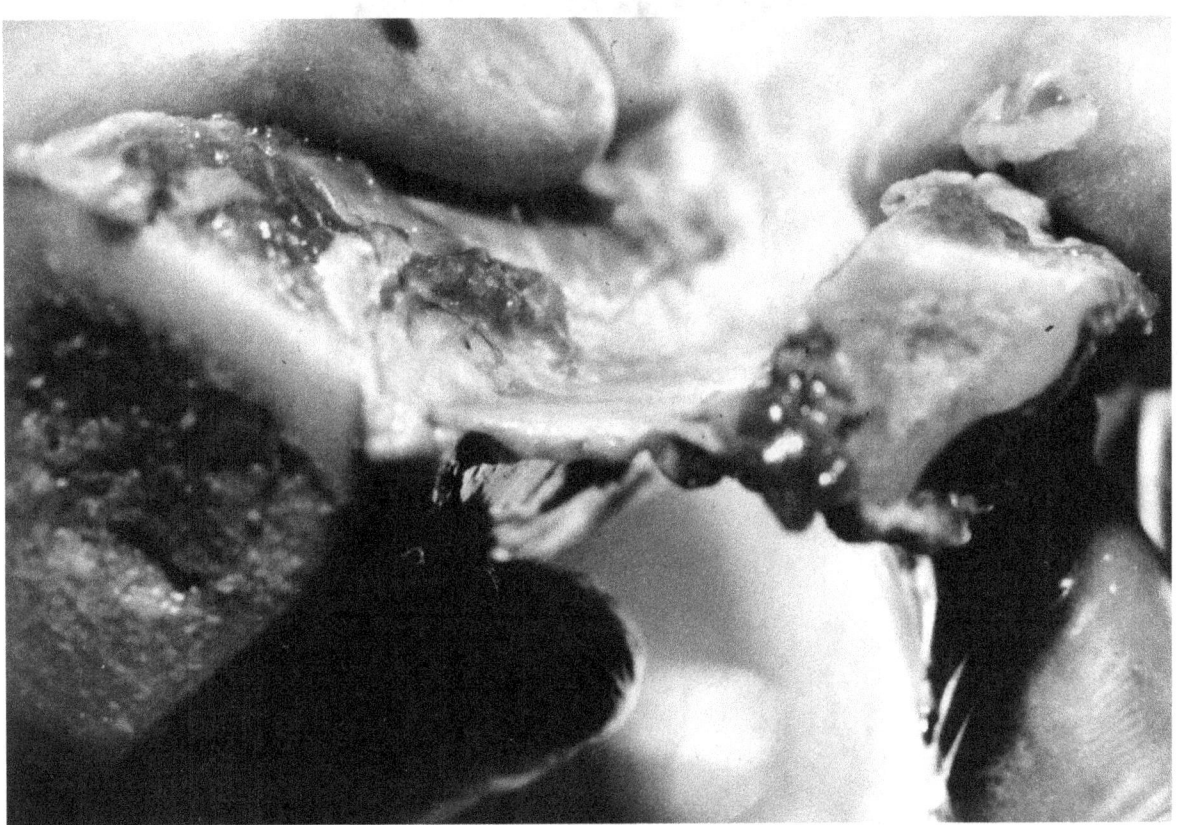

Fig 30 A

Fig 30 B.

Fig 30.

End view of tibia, fibula and interosseous membrane (A). anterior view of interosseous membrane with muscle attachment reflected back (B). Note, shiny, highly figrous structure.

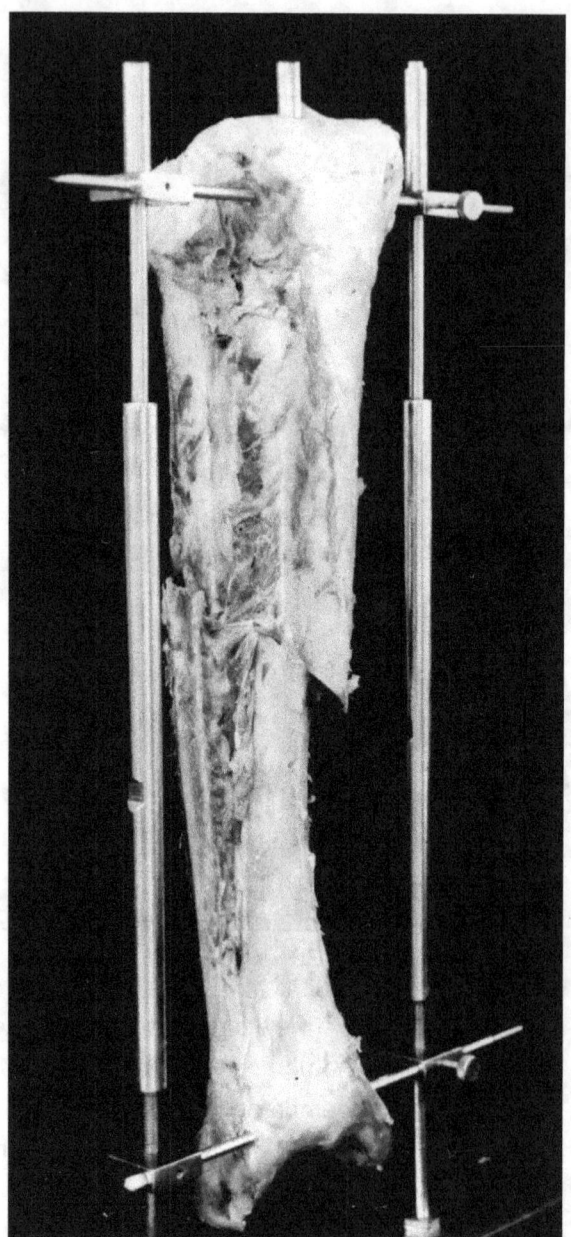

Fig 31. Vertical load applied to specimen in fixture which

allows no angulation in varus or valgus

It appears that if the damage to the membrane is small, only a minor contribution from other sources (such as a fracture brace) is necessary to stabilize the limb against angulatory deformities (Fig. 33).

Fatigue could be important in creating damage in the membrane and periosteum at the fracture site. _The specimens tested were loaded statically with relatively small loads (28 pounds) and negligible shortening and angulation took place if the interosseous membrane and periosteum were intact. But if the loads were cycled or the fractures were manipulated under load, creep took place in the membrane without accompanying deformities at the fracture site. Specimens were fixed in formalin and therefore, no conclusions of quantitative significance can be drawn from these studies.

In summary, the role of the interosseous membrane is complex and may be influenced by such factors as orientation of the fracture and displacement and angulation experienced at the time of
the initial injury.

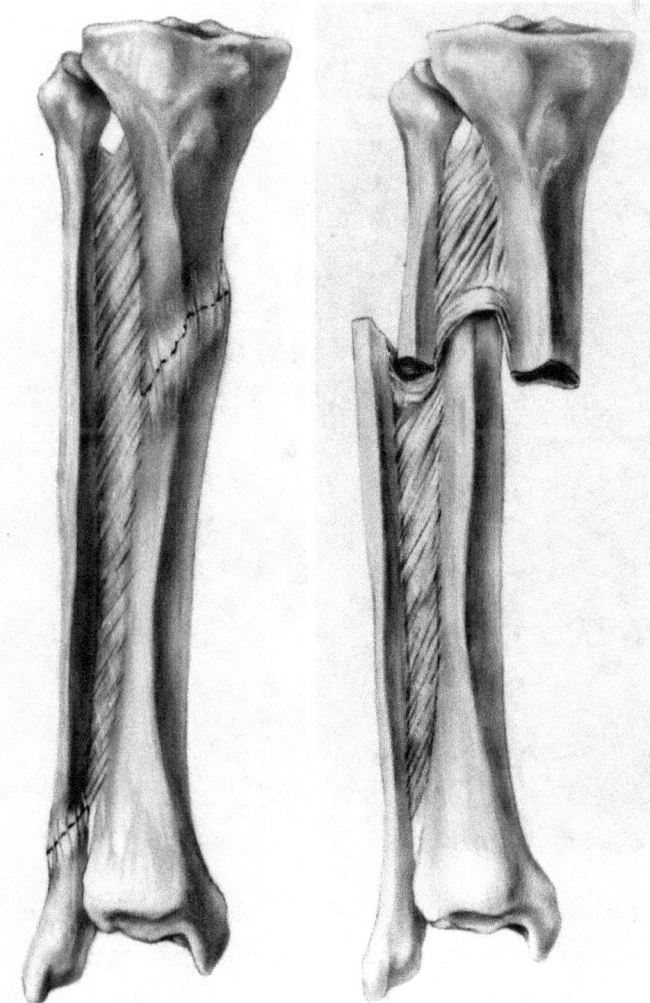

Fig 31

Perisotium and interosseous membrane holding fractures at different levels of tibia and fibula (A) and at the same level (B). Note the difference of amount of membrane holding fragments and the effect of direction of displacement due to orientation of the fibers of the membrane.

Fig 31 A   Fig 32 B

Fig 32 C

Fig 32 D

Fig 32 C-D. Damage due to angulation of fractures at different levels (C) and at the same level (D) demonstrate the possible effects of initial displacement at the time of injury. In one case damage must be extensive (C); the other (D), not necessarily so. Thus, the more stable case (A) becomes the least stable if initial displacement in extensive (C). However, the less stable case (B) is potentially more stable in the face of extensive initial displacement.

Fig 33 A.

Figure 33

If interosseous membrane in intact when angulation occurs (a) and (b), fragments could be maintained in adequate alignment under axial load for reduction with minimal counter moment (as applied by thumb) in (c).

Thus if interosseous membrane is intact, a small contribution by the brace could easily counteract the moment applied by the external load and maintain angulatory stability

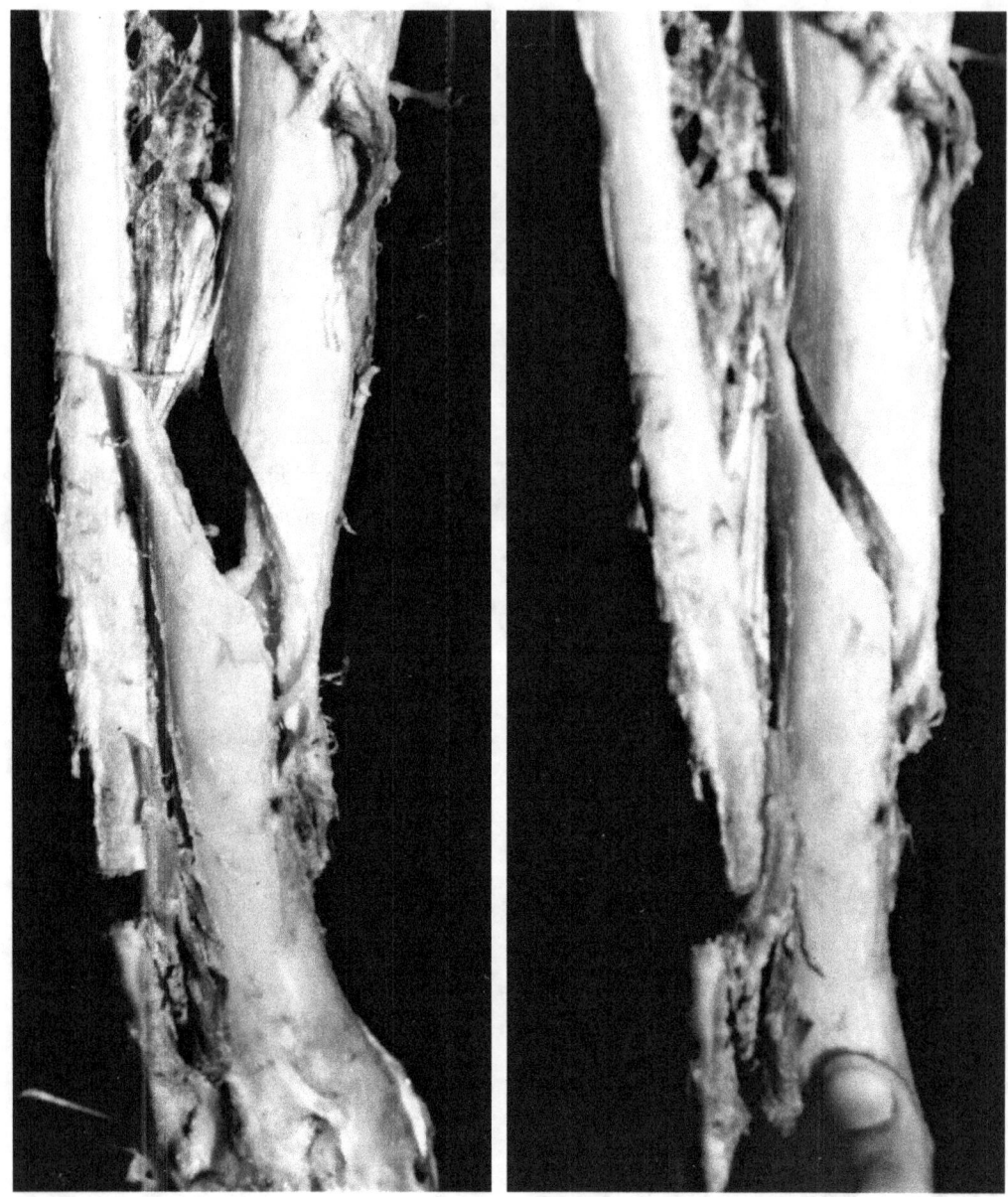

Fig 33 B.   Fig 33 C.

EXPERIMENT 5:

Failure Mode and Properties of the Interosseous Membrane

In the previous experiment, the possible contribution of the interosseous membrane to fracture stability was studied. Those studies (since they have no quantitative significance) rest on the assumption that the mechanical properties of the interosseous membrane are significant. The following investigations deal with the structure of the interosseous membrane and its interaction with surrounding tissues.

Procedures

Three different experiments were performed. The first consisted of macroscopic observations of the interosseous , membrane during traction with-and across the grain of its fibers (Fig. 34). The second consisted of microscopic observations of the same phenomenon in order to closely observe its mode of failure. Some specimens had been preserved in formalin and others came from freshly amputated limbs.

The third experiment consisted of tensile tests of specimens on an Instron tensile machine with recordings of load-deflection characteristics (Fig. 35). The mechanical experiments were performed in a sealed container filled with saline solution at room temperature.

Fig 34 A

Fib 34 B

Fig 34

A macroscopic view of interosseous membrane failing under loads applied perpendicular to the axes of the tibia and fibula.

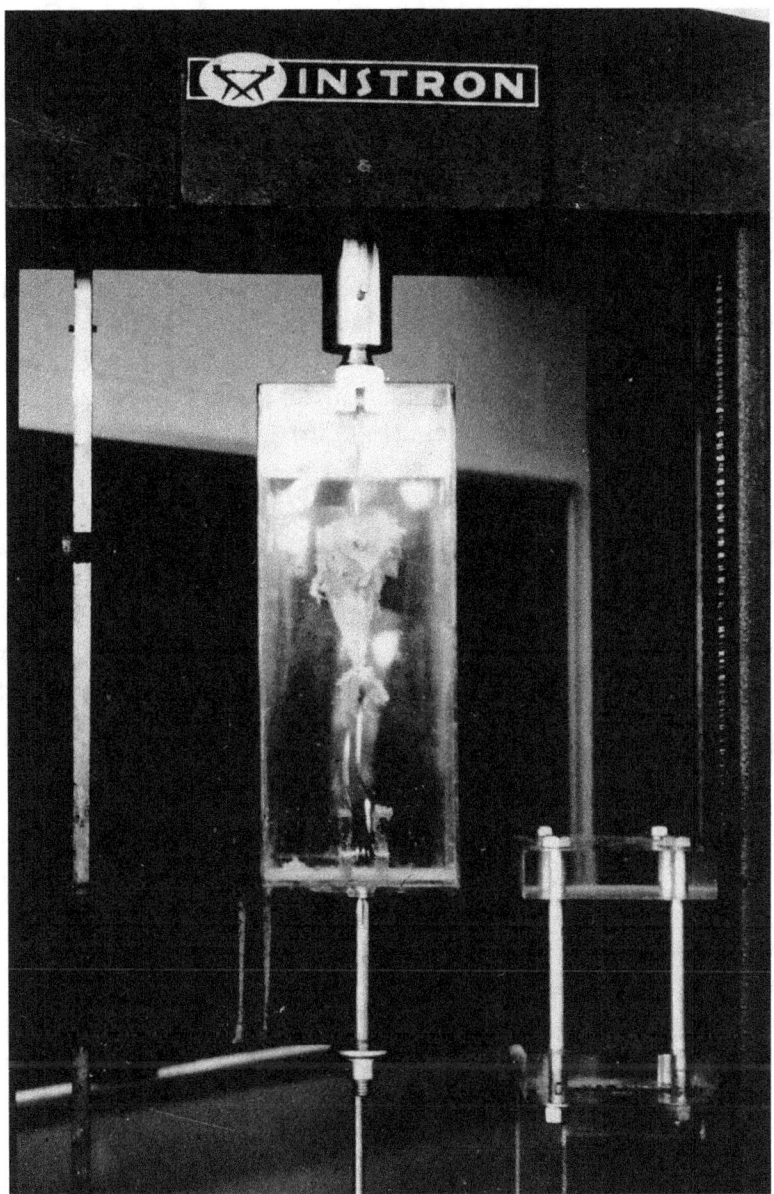

Figure 35 – Tensile testing machine with fresh specimen of interosseous membrane in fixture which allows testing at room temperature in saline solution of whole sections of tibia, fibula, interosseous membrane and periosteum.

Results and Discussion

The interosseous membrane is a connective tissue composed primarily of fibers of collagen rather than a protein-phospholipid sandwich layered membrane. It has a very regular, unidirectionally oriented fibrous structure which bridges between the tibia and fibula and interdigitates with the periosteum and adjacent bone collagen. Muscles attach to the periosteum and the interosseous membrane in numerous places and vessels perforate it in several places at its proximal and distal ends. The interosseous membrane, like most biological tissues, is a viscoelastic (probably non-linear) material, anisotropic and non-homogeneous. It appears to behave as a composite system and might also be expected to behave in a manner somewhat similar to an elastomer with semi-permanent plastic behavior [13].

Connective tissues in general exhibit non-linear elastic behavior at low strain. For such tissues the non-linear initial portion of the curve generally represents a rearrangement of the fibers as evidenced by polarized light microscopy [2,9,68]. The rearrangement in the case of tissues made up of fibers held together by ground substance, with cells and vessels intermingled (a composite system), is a reorientation of the fibers which places a "preload" on the matrix. This type of strain serves to take the "slack" out of the*system, but if the process does not cause failure in the matrix, the slack can be recoverable [9,68,71]. Once the slack is taken out of the system (elastically or plastically) it becomes much stiffer and the fibers are loaded through the remaining ground substance composite system (Fig. 36).

When the interosseous membrane is loaded in tension parallel to a the axis of its fibers, it behaves much like tendinous structures which also have unidirectional orientation of their fibers [9,68]. However, if the membrane is loaded perpendicular to the axis of its fibers, the system pulls apart like a piece of expanded metal (Fig. 37). The fibers group into strands which attach to other strands at relatively regular intervals forming hinged nodal points. These connections between strands bend to form a series of parallelograms in a matrix of its own. Since the nodes of the matrix act like hinges, the system can expand by simply changing the angles in the parallelograms so that they fold back on themselves to allow expansions of the system. In this way the system can accomplish a great amount of elongation without incurring extensive damage to the matrix. After the "slack" of this system has been taken out (plastically), the fibers, in a" folded weave pattern, begin to bear the load and the membrane deforms elastically until a peak load is reached. After the peak load, the system fails plastically. Once the initial expansion has taken place the behavior of the load deflection curve for a mode of failure perpendicular to the grain of the fibers, is almost exactly like the load deflection curve for a mode of failure parallel to the grain. The peak load borne by the system in the perpendicular mode is only about 60% or less of the peak load borne in the parallel mode. The slope of the elastic portion is lower than in the parallel mode. This occurs because failure is taking place in the matrix during the initial expansion in the perpendicular mode., when it finally goes into the phase of elastic behavior there is less material remaining to hold the fibers together (Fig. 37).

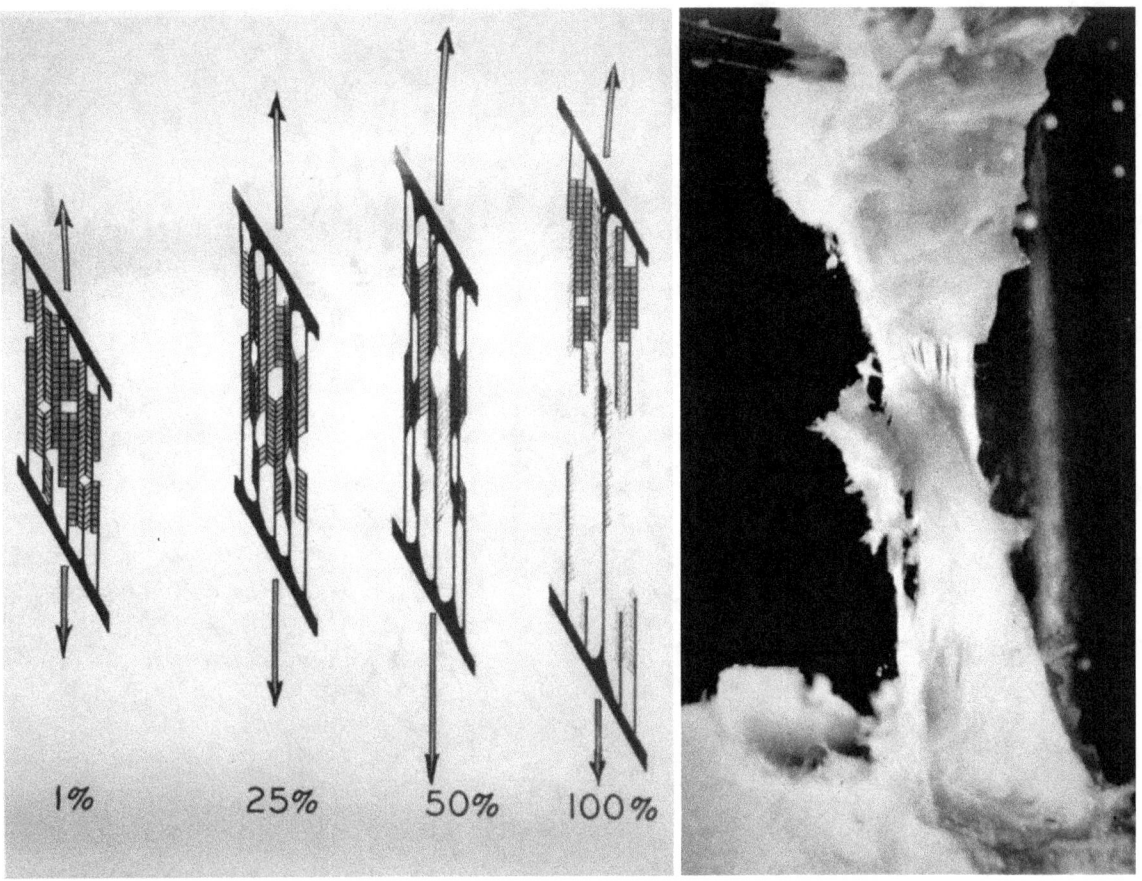

Figure 36 – Model of failure mode of interosseous membrane loaded parallel to its grain. Initially most matrix (represented by horizontal lines) is relaxed. As some matrix deforms elastically, the non-linear portion of he load deflection curve begins and the slope steepens as more matrix begins to deform. At 10 to 25% elongation, plastic deformation begins and some failure takes place in the matrix, but the elastic behavior dominates. At 50% elongation the peak load is met and almost everything that has not failed is in plastic deformation. Fibers group into bundles, and failure progresses until the last strand lets go at low load but about 100% elongation. (A) Gross specimen in parallel mode of failure (b) during tensile test.

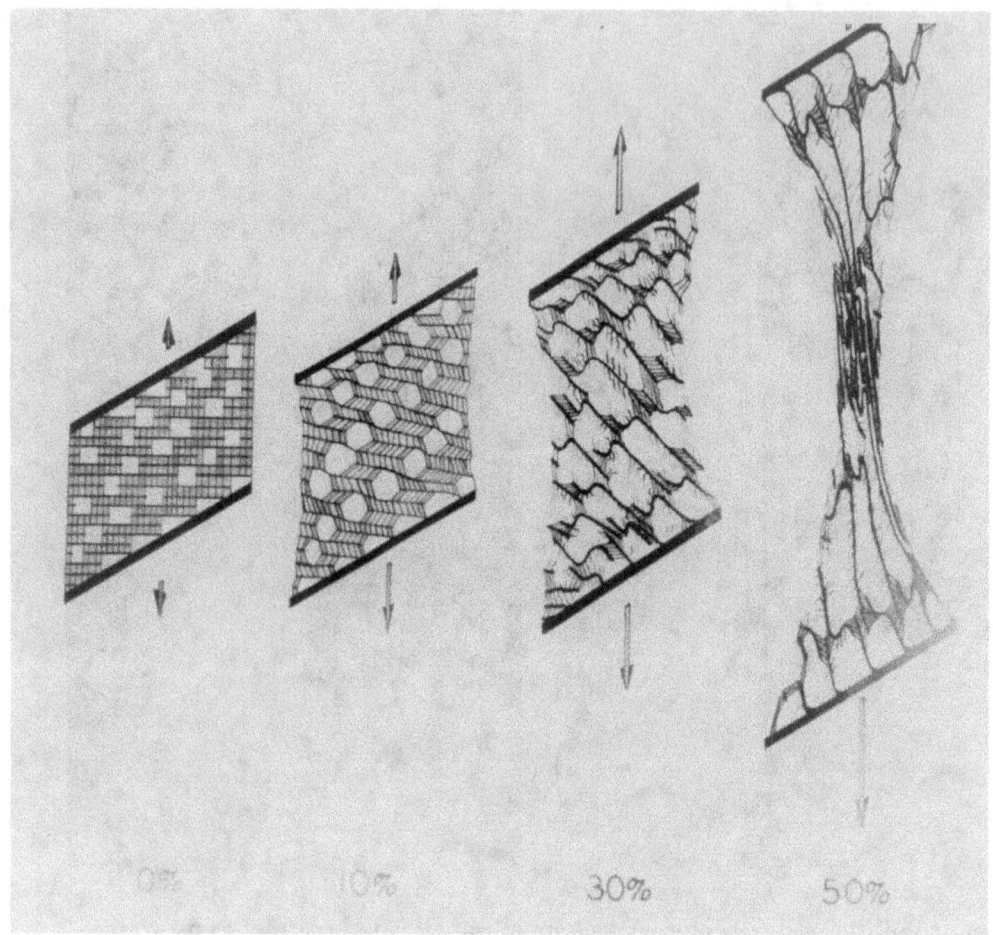

Figure 37 – Model of failure mode of interosseous membrane perpendicular to its grain. The matrix id elastically deformed almost all at once initially with small amounts of plastic deformation near the holes or discontinuities due to stress and load concentration/ as holes open up failure of small amounts of the matrix occurs and other small amounts plastically deform near the enlarged holes. Elongation is increased without much increase in the load as holes become parallelograms with fibers as the sides and matrix holding the fibers together at periodic locations forming hinges for the parrelograms to open from. As the hinges fold back on themselves, the fibers realign perpendicular to the original direction. The failure at peak load seems to be similar to the failure model parallel to the grain, except for the gradual sweep from the wide portion of the specimen to the narrow region of failure. A gradient of parallelograms of fibers apparently form the hyperbolic like curve from the widest to the narrowest portion of the specimen.

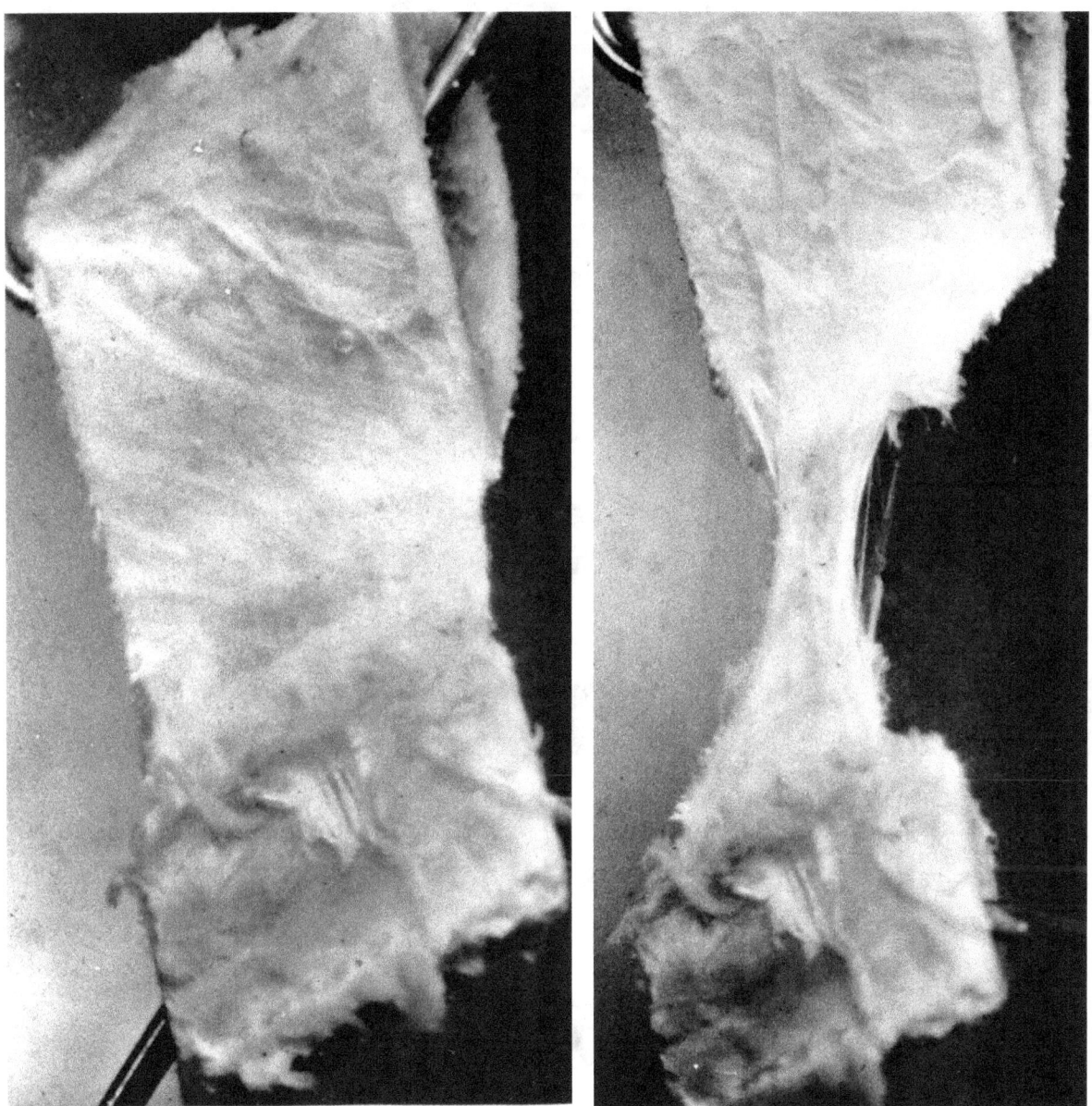

Figure 37 – Specimens of interosseous membrane with periosteal attachment to a portion of tibia and fibula under tensile test in perpendicular mode of failure, with minimal elongation (b) and extensive elongation (c).

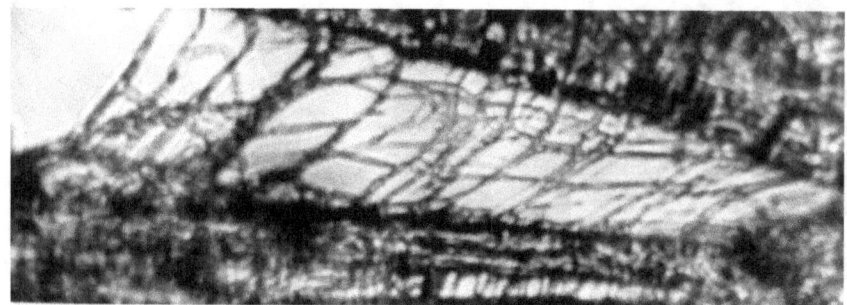

Fig 37 D

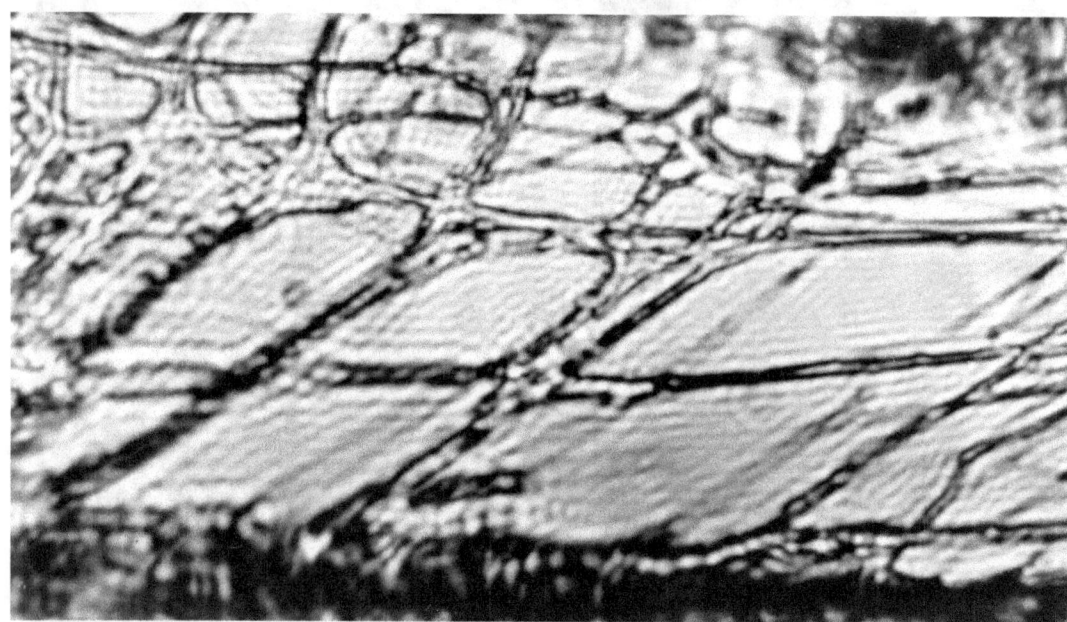

Fig 37 E

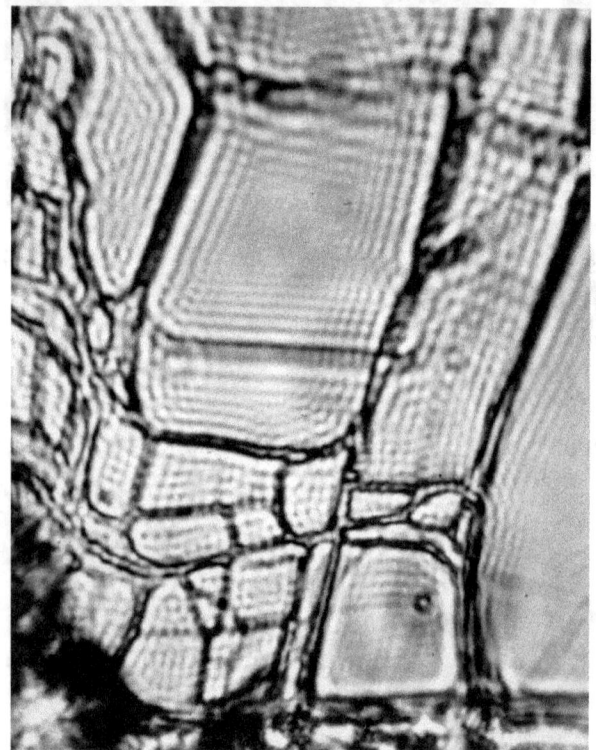

Figure 37 – Microscopic view of fibers expanding under load perpendicular to the fiber direction (d).

Close up of the parallelograms beginning to open in the region of failure (e) As failure progresses, the hinged nodes allow the parallelograms to open more allowing elongation without extensive failure in the matrix.

The mode of failure under cross grain load is a combination of both perpendicular and parallel failure modes. First some areas begin to fail in the perpendicular mode since it offers the least resistance. This allows large sections between the areas of perpendicular failure to reorient themselves parallel to the direction of load. Then after the slack is out of these systems,-the resistance to load increases sharply as the fibers parallel to the load dominate the load deflection curve.

Conclusions and Speculations in Reference to Clinical Applications

The interosseous membrane can elongate approximately 120% in tension along the grain of the fibers before failure. More importantly, the membrane does not reach its peak load carrying capability until it stretches about 50% (Fig. 38). Across the grain the membrane can take less load but it can elongate by more than 300% before failure and about 100% before it reaches its peak load carrying capability. The strain damage one might expect in a typical injury would probably be primarily cross grain to the membrane (i.e. separation of the tibia and fibula, Fig. 39) which can be significant before failure. After the injury it seems that stability from the membrane comes primarily from tension along the grain of the fibers. Thus injury to the membrane could be significant before the stability which the membrane can provide is seriously jeopardized. This could partially explain the clinical observation that fractures of the tibia do not shorten beyond the initial shortening experienced at the time of injury.

The interosseous membrane demonstrates sufficient strength to be effective in stabilizing fractures, at least for short duration loads. But the interosseous membrane also demonstrated a rapid stress relaxation response which indicates a tendency for low creep resistance. Static loading conditions might produce additional shortening due to creep in the interosseous membrane if other factors and systems were not present to prevent it.

EXPERIMENT 6:

Soft Tissue Damage in Humans

Since it appears that damage to the membrane can be extensive before failure occurs, the amount of actual damage to the interosseous membrane was studied in actual fracture victims.

Procedure

Accident victims who sustained tibial and fibular fractures and died from other causes were studied. Before the dissection of the fractured limbs, x—rays were taken with and without loading (Fig. 40)

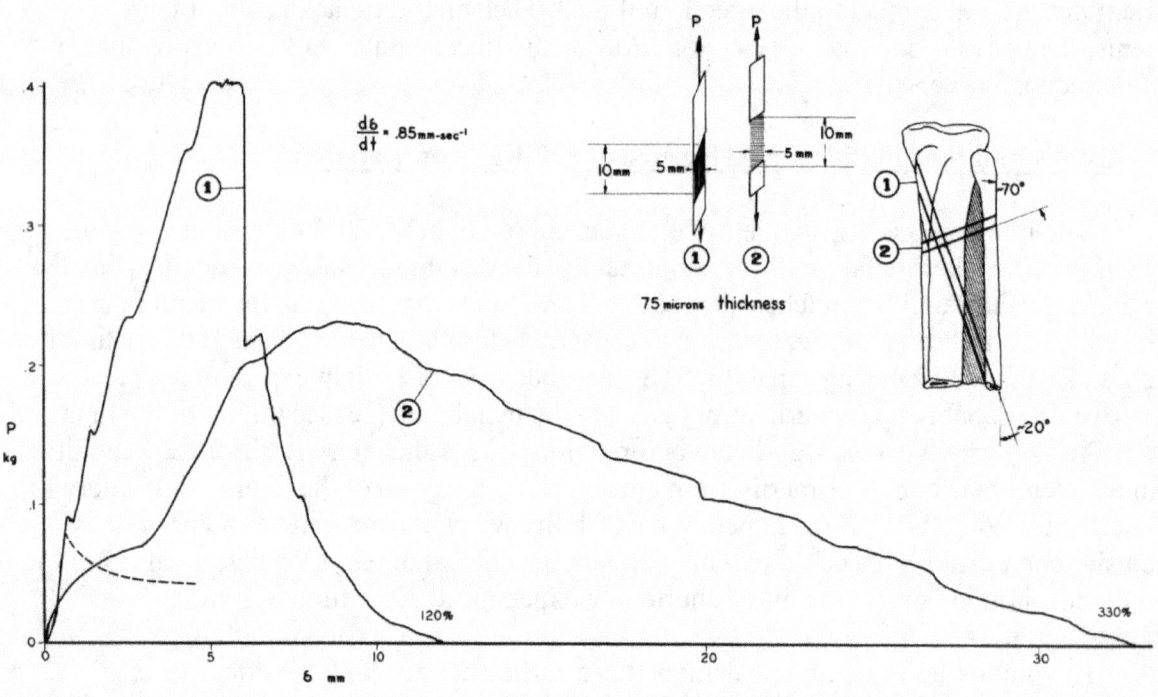

Figure 38 – Typical load-lengthening curve one might expect from two specimens of the dimensions shown in the right upper quadrant indicating tensile specimen sites and dimensions.
   (1) Is parallel to the fiber direction and (2) is perpendicular to fiber direction. Note greater
        load in (1) but greater elongation in (2).
Note apparent elastic deformations particularity in (1). Notches in the elastic portion of the curve indicate failure of small groups of fibers at various amounts of straqin before the ultimate strength of the membrane is reached. Inelastic initial deflection in (2) is due to the "expanded metal" effect as noted in figure 37.

Observations and Conclusions

        Observations of four fractures were recorded in deceased humans It was interesting to note that even in the face of extensive damage once the fragments reached their apparent initial displacement the soft tissues stabilized the fracture in that position (Fig. 41). The soft tissues attached to the interosseous membrane are important in providing stability if the membrane is severely damaged. Therefore, it is logical to assume that the interosseous membrane is less effective by itself than it is in conjunction with the other soft tissues. The behavior of these post—mortem specimens indicated great similarity to the artificially created fractures. It appears that our laboratory models of tibial fractures in fresh above-the-knee amputations are reasonable approximations of in vivo fractures.

EXPERIMENT 7:

Mechanical Response of Tibial Fractures in vivo

Previously we investigated the mechanical response of dead tissues on the stabilization of tibial fractures created in vitro. The active properties of live tissues were studied in the following investigations.

Live patients within 24 to 48 hours after having sustained fresh oblique tibial fractures with associated fibular fractures were observed under fluoroscopy. Small loads were placed on the limbs in axial compression, bending and torsion, similar to the loading conditions used on limbs in the laboratory. The resultant motion at the fracture site was recorded by cine radiography. The loads were applied with and without a fracture brace.

Results and Conclusions

Four patients were tested. The responses in all cases were similar. The fragments displaced under loading conditions with and without a brace, but the displacements were greater in the non-stabilized extremities. In torsion there was better stability in the braced limbs, but no significant differences were observed.

In all instances the displacements were fully recoverable (elastic) upon relaxation of the load. The only difference between the response in and out of the brace was the magnitude of the displacements. The recovery of the initial shortening and alignment , after relaxation of load was consistently rapid and complete (Fig. 42).

Discussion

The rapid return of the fragments to their initial position upon relaxation of loading may be due in part or in whole to reflex stimulated muscle activity. ~This would constitute an active property of the muscle which has not been duplicated in the laboratory model.

78

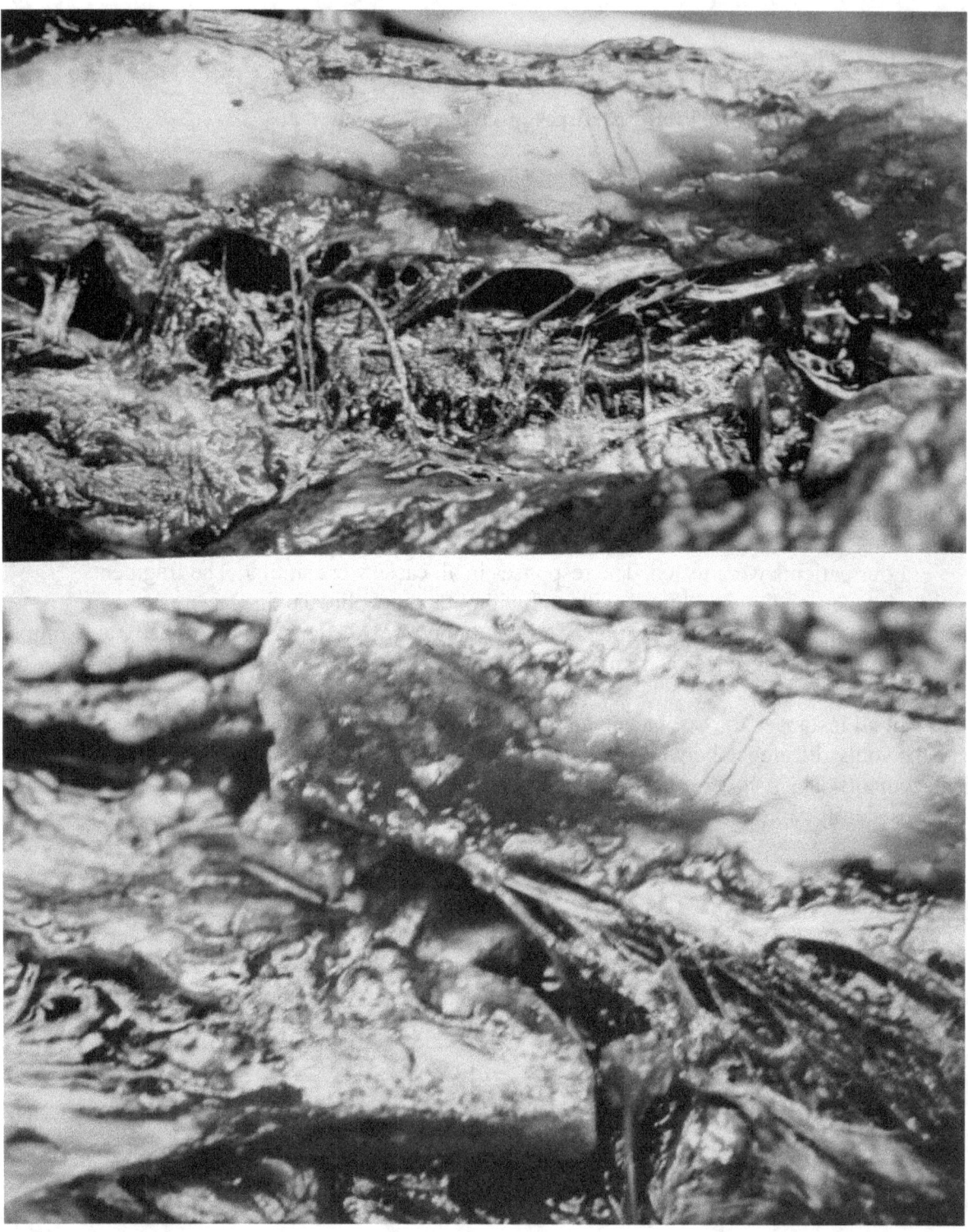

Figure 41 – Lateral view of fracture tibia with exposure of interosseous membrane. Note shredding of membrane and disruption of soft tissues as expected for fractures at different level with extensive initial displacement and gross instability.

EXPERIMENT 8:

Dynamic, Passive in vitro Modeling

The following experiment was conducted in an effort to develop a dynamic, passive laboratory model which would approximate the observations made in clinical practice and in vivo studies. As noted in the previous studies of actual soft tissue damage, isolation of the role of a single soft tissue structure was probably an over simplification of a complex interaction of soft tissues and skeletal structures. Thus another model was developed to study the mechanisms by which soft tissues can further stabilize a tibial fracture.

Procedure

A first model consisted of fresh above-the-knee amputation specimens to which cycled loads were applied and the effects recorded by cine radiography.

☐   A second model was developed to observe displacement of the soft tissues during dynamic displacement of the fractured bones. This second model consisted of fresh, above-the-knee amputations cut in a sagittal plane. The half limbs were placed in hemi-braces with a flat, clear, acrylic window on the side of the sagittal cut (Fig. 43). Loads were placed on the specimens in cyclic axial compression. Resultant dynamic deformations were recorded on 16 mm color film.

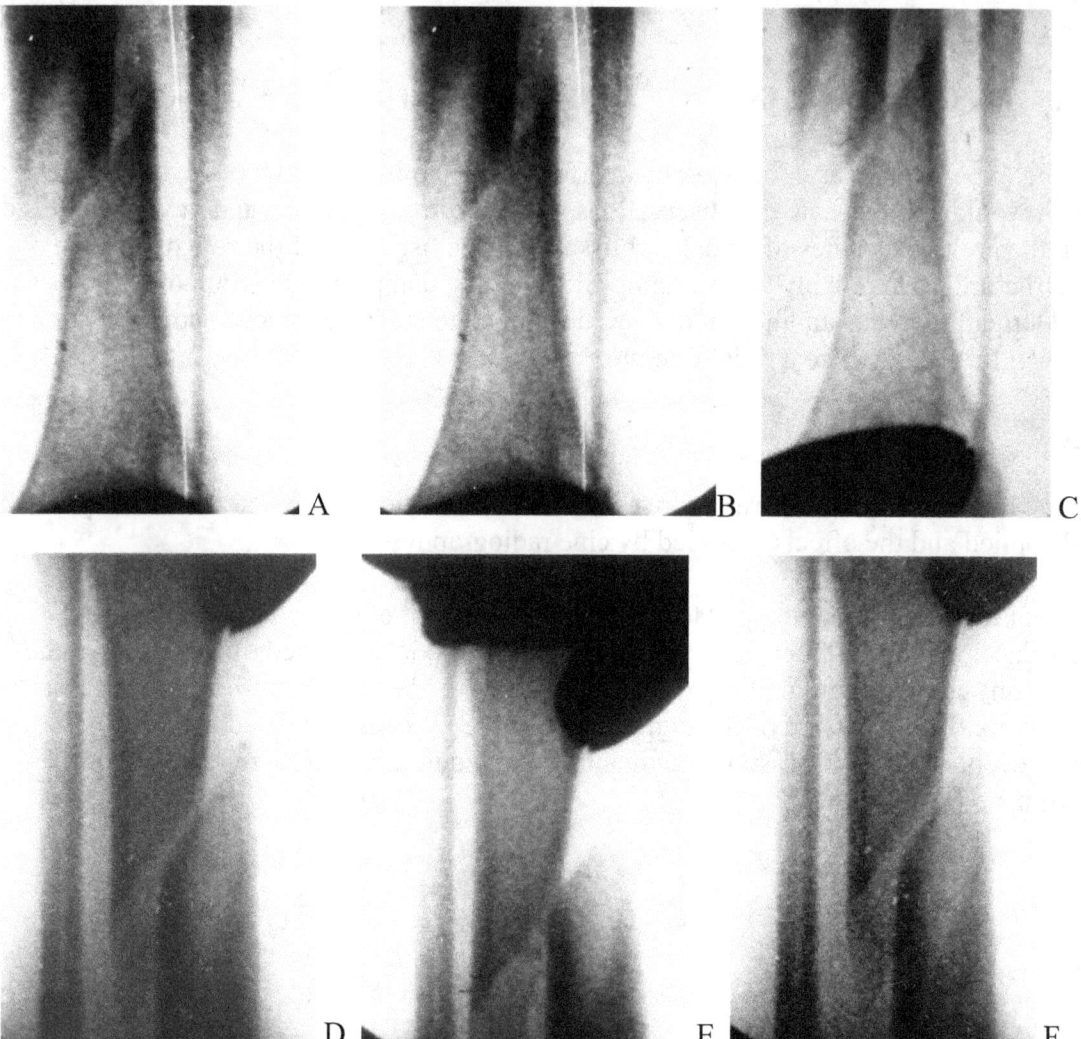

Figure 42 – Summary of cineradiographic recording of motion at the fracture site. With the brace applied to position of the fragments before load (a). with load applied (b), and upon relaxation of load (c). Without the brace the position of the fragments before load applied (d), with the load applied (e) and upon relaxation of load (f). Note that displacements from these loads are completely recoverable upon relaxation of load (elastic)

Results and Conclusions

The limbs studied by cine radiography responded to loading conditions in a manner similar to that of limbs tested passively in vivo. The deformations under small loads with and without a brace were fully and rapidly recoverable upon relaxation of load. Deformations in axial compression and bending were most pronounced without the brace and were reduced in magnitude when stabilized in the brace. Rotational displacements were controlled by the brace but not as significantly as angulatory and shortening deformities. The most significant difference between braced and unbraced limbs under rotational loads was the rate of return of the fragments to their initial position. These observations correlated well with the observations of passive in vivo behavior under identical low level loading conditions.

Since the laboratory model seemed to be a reasonable approximation of the in vivo condition, we extended the loading to levels too high for experimental application to live patients. Without the brace, the limbs demonstrated poor angulatory stability under high loads (up to 450 pounds compression and 15 feet pounds bending) but the soft tissues only yielded slightly to shortening. The deformities that developed from high loads applied without the brace were not encountered when similar loads were applied to the braced limbs.

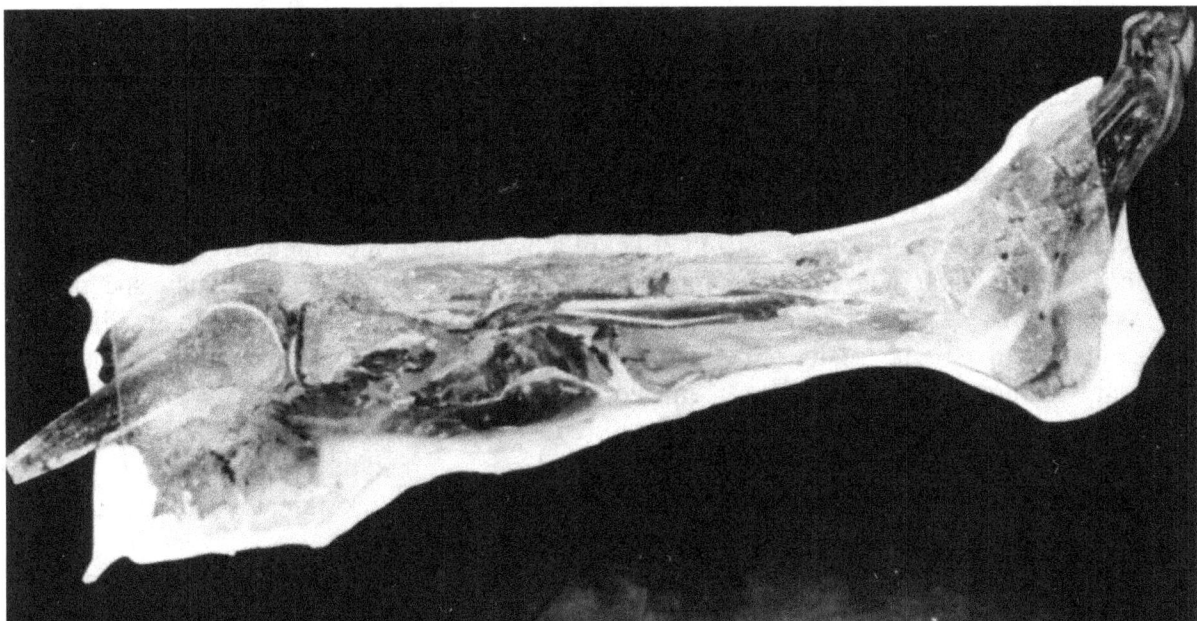

Fig 43 A

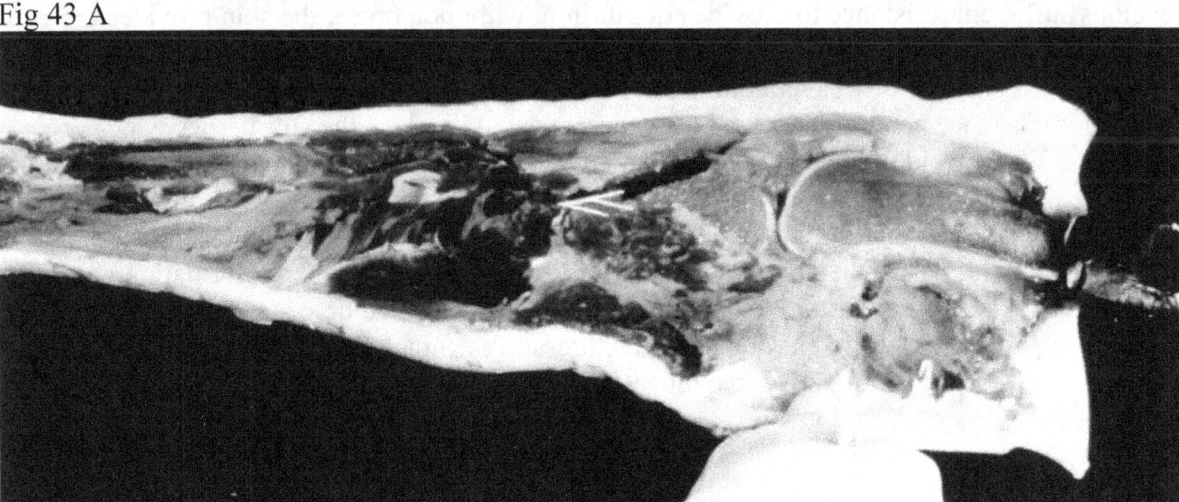

Fig 43 B

Figure 43 – View of a fresh A-K amputation after having been cut through a sagittal plane and mounted in an orthoplast (half brace) with an acrylic window along the sagittal plane (A) as loads are applied to the systems through the femur, motion at the fracture site is recorded by 16 millimeter movie, and marks indicting the excursion of the fragments are made on the window indication by the white arrows (b)

☐   We concluded that the soft tissues in their passive state provide an elastic support to the fragments and in conjunction with a brace can prevent deformity even under high loads. This system probably allows small elastic motions to take place at the fracture site, at least before the development of intrinsic stability.

The "hemi-legs" (the second model) demonstrated similar elastic motion at the fracture site with cyclic loads. It was seen that the soft tissues moved posteriorly and distally during movements of the bony fragments. In some cases the loads were transferred completely through the soft tissues from the proximal to the distal fragment without direct interaction between them. Therefore, it appears that anatomic reduction of a fracture is not a prerequisite for adequate stability.

No backward movement of the soft tissues was observed as would be expected if the mobility of the tissues was sufficient to give pure hydraulic action as speculated earlier. Instead, there appeared to be a compartmentalized effect of a hydraulic or incompressible fluid nature". The muscle tissue did not appear to supply the elastic response as we had previously speculated 54, it simply provided the hydraulic response. The muscle in its compartments separated by walls of connective tissue, seems to deform into a compliant manner to fit the shape of the adjacent structures. The outer compartments conformed to the walls of the brace and the inner compartments to the outer ones, etc., until pressure developed within the tissues. Upon relaxation of load, the elastic connective tissue boundaries returned the muscle compartments to their original shape and thus the fragments to their original position.

Therefore, if the outer boundary condition (the brace) is compliant or of poor fit, it allows the soft tissues to deform further until they fill the gaps in the system or stretch it until it develops sufficient resistance to provide equilibrium. without a brace, the skin provides the only outer boundary and thus the boundary becomes more compliant than with the relatively rigid brace. Regardless of what condition causes the extra "slack" in the system the result is the same. More displacement of the fragments is necessary to displace the soft tissues which are required to fill the void of a poorly fit brace or stretch the compliant boundary of the skin.

There is a practical limit to the degree to which this phenomenon can take place. The soft tissue compartments can only deform to a limited degree before their deformation provides resistance by itself sufficient for equilibrium of the system. Thus, a brace or a leg with a hole in it will not allow the tissues to extrude under the pressures of internal loads applied to the soft tissues by the fragments. It is this same intrinsic integrity of the soft tissues that does not allow the tissues to extrude across the joints or move great distances within the leg. This integrity can be demonstrated empirically if one thinks about attempting to push a solid rod through a piece of raw beef steak. The tissues will displace to some degree, but soon the system resists very strongly.

Through their inherent intrinsic strength the soft tissues of the leg, if intact or damaged minimally, can stabilize the fragments of a tibial fracture. with the aid of a relatively rigid, well fit brace the stability can be improved considerably, reducing the amount of motion at the fracture site. .

⬜

The limitations of these models must be kept in mind whenever conclusions are to be drawn from experiments utilizing them. The whole leg model does not duplicate the internally applied loads off active, voluntary muscles. The laboratory studies define the limits of displacement and how these limits can be controlled. In the "hemi-leg" models, the motion is limited to two dimensional movements due to the introduction of the "window". This window also permits a plane of negligible shear stress resistance and high compression resistance which is not present in the whole leg. But the resultant movements of the fragments to loads placed on them in the plane of motion allowed by the model duplicates they motions in whole legs closely enough to be a reasonable approximation for the study of related soft tissue displacements.

In summary, we have established two laboratory models for the study of passive stabilization of tibial fractures which appear to be appropriate for the study of stabilization of fresh fractures and to aid in the development of treatment techniques. These models are not quantitative at this time but our findings may be correlated to clinical studies 34.

⬜

EXPERIMENT 9:

The Effect of Soft Tissue Damage on Stability, In Vitro

In the previously described studies, the importance of soft tissue 'damage was studied in context with its probable role in providing stability to tibial shaft fractures. The importance of soft tissue damage has been mentioned by others 3'5'11, in relation to prognosis and management. Their role in stability, however, is not fully understood. The "hemi-leg" model described in the previous chapter provided some information concerning their possible role in stabilization.

Procedure

Fresh above-the-knee amputation specimens were cut along a sagittal plane. The half legs were fractured and mounted in braces with transparent acrylic windows over the fracture sites. Loads were applied and the displacements visually observed and photographed. The soft tissues were then selectively cut in various degrees attempting to create progressive instability.

The same experiment was repeated in whole legs recording the findings by cineradiography. The effects of full, single, double and triple compartment fasciotomy and fibulectomy were studied singly and in combination.

Results

Under vertical loading of fractured tibial specimens no appreciable difference was observed between specimens without soft tissue damage and those with damage localized at the fracture. If full integrity of the soft tissue across the joints was present the loads transferred well through the soft tissues from the proximal to the distal fragments. As damage was created across the various muscle compartments in the posterior calf area, mobility of the tissues seemed to increase. and backward extrusion of material began to take place toward the knee joint. The displacement stopped as the fragments "settled into" a new position which seemed to take up the "slack" on the system created by increased soft tissue damage. Once this newly created slack was gone, and the new additional shortening took place, the fragments displaced minimally and elastically about their new position. Further damage, increased the "slack" and subsequent shortening but the fragments seemed stable in their new position each time (Fig. 44). It was virtually impossible to destroy the tissues sufficiently to allow complete mobility of the tissues and the bones within them. we felt that all the soft tissues would have to be dissected, chopped and poured back into the leg to be able to create full mobility of the bone fragments through the tissues.

Following fasciotomy, gross instability in bending, axial compression and torsion were observed when the experiment was carried out without the brace. However, with the brace, the limb showed only minor or neglible increases in the displacement of the fragments. Similar findings were noted following fibulectomies. The fragments became so unstable without a brace that it was impractical to load them as in previous studies. However, after reduction in the brace, the system became stable (Fig. 45).

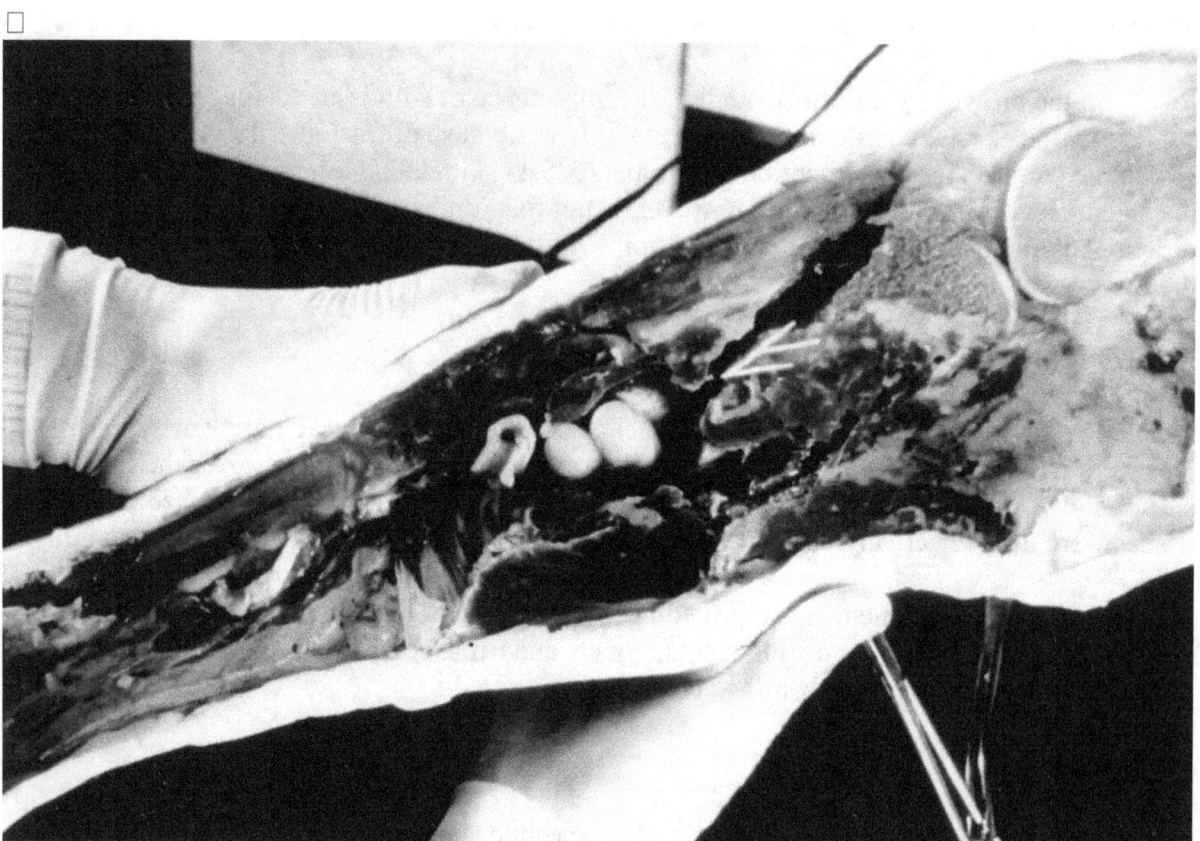

Figure 44A

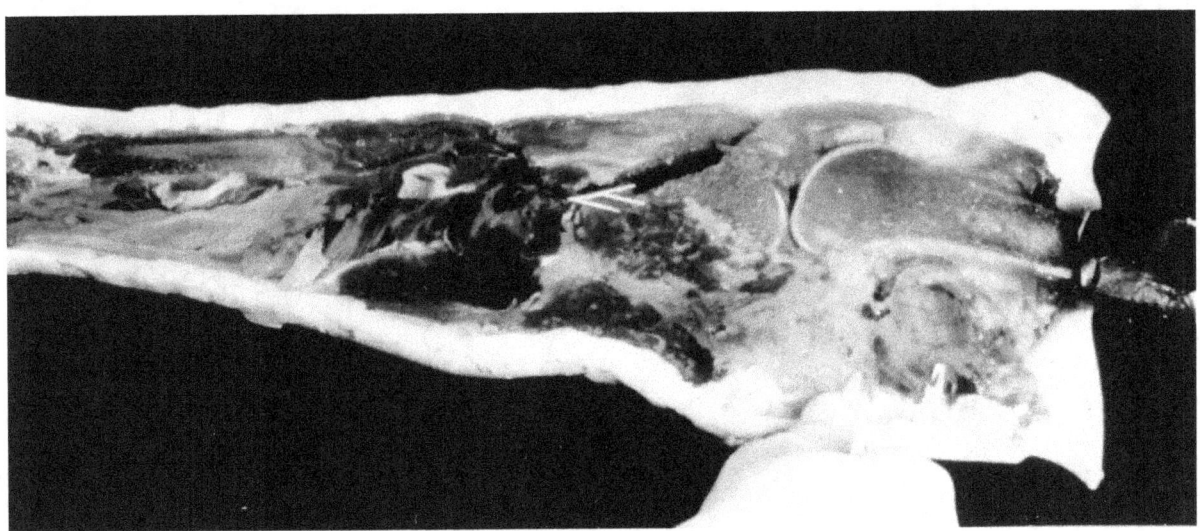

Figure 44 B

Even in the face of extensive soft tissue damage demonstrated by the figures protruding through the leg, (a) the fragments maintain approximately their original position and range of motion, as long as the fit of the brace was maintained.

When the fit of the brace was lost, however, by splitting the brace and peeling it from the leg, the fragments were allowed to shorten and collapse into the damaged soft tissues when load was applied (b)

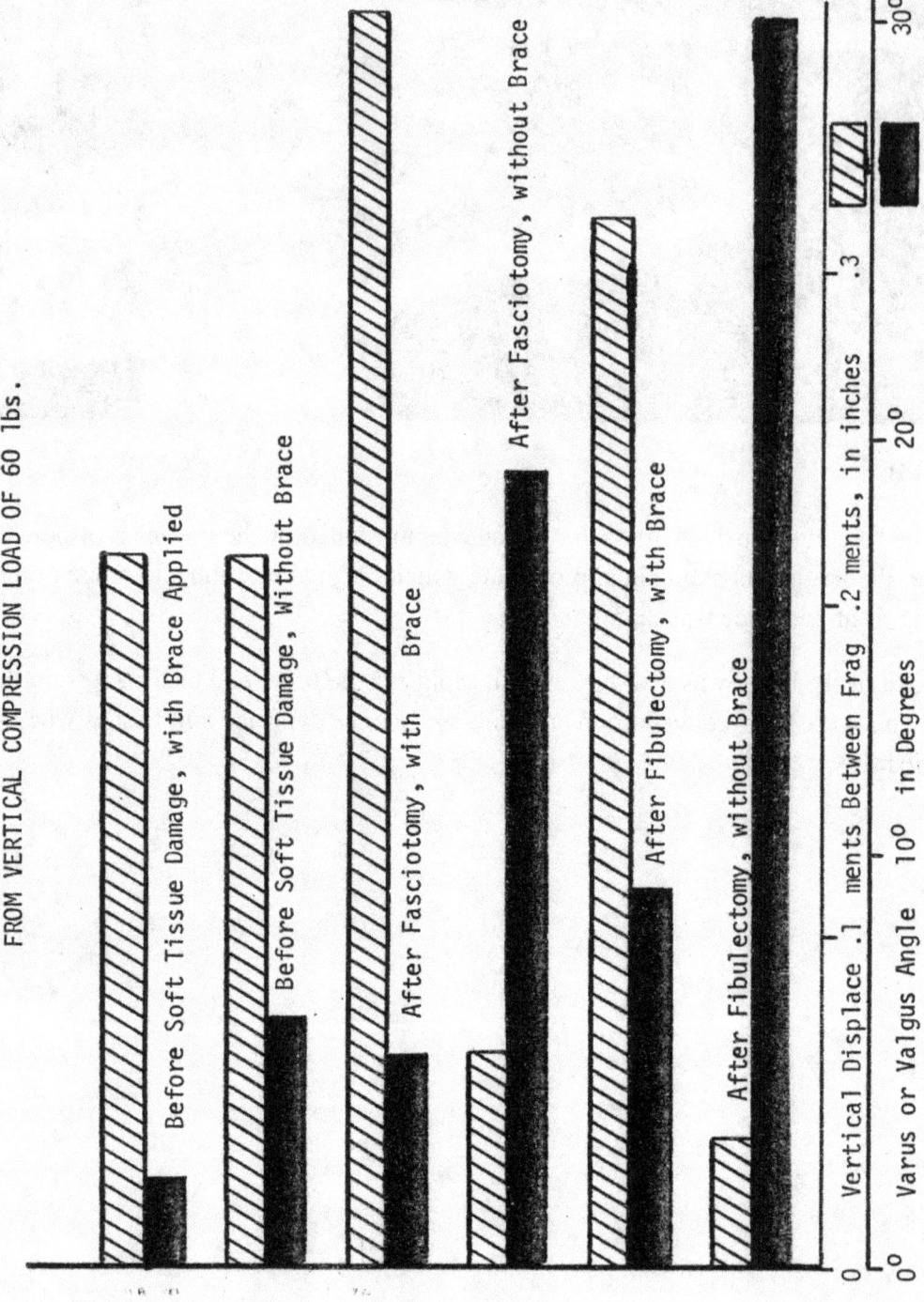

Figure 45a – Elastic Motions measured at the Fracture Site.

87

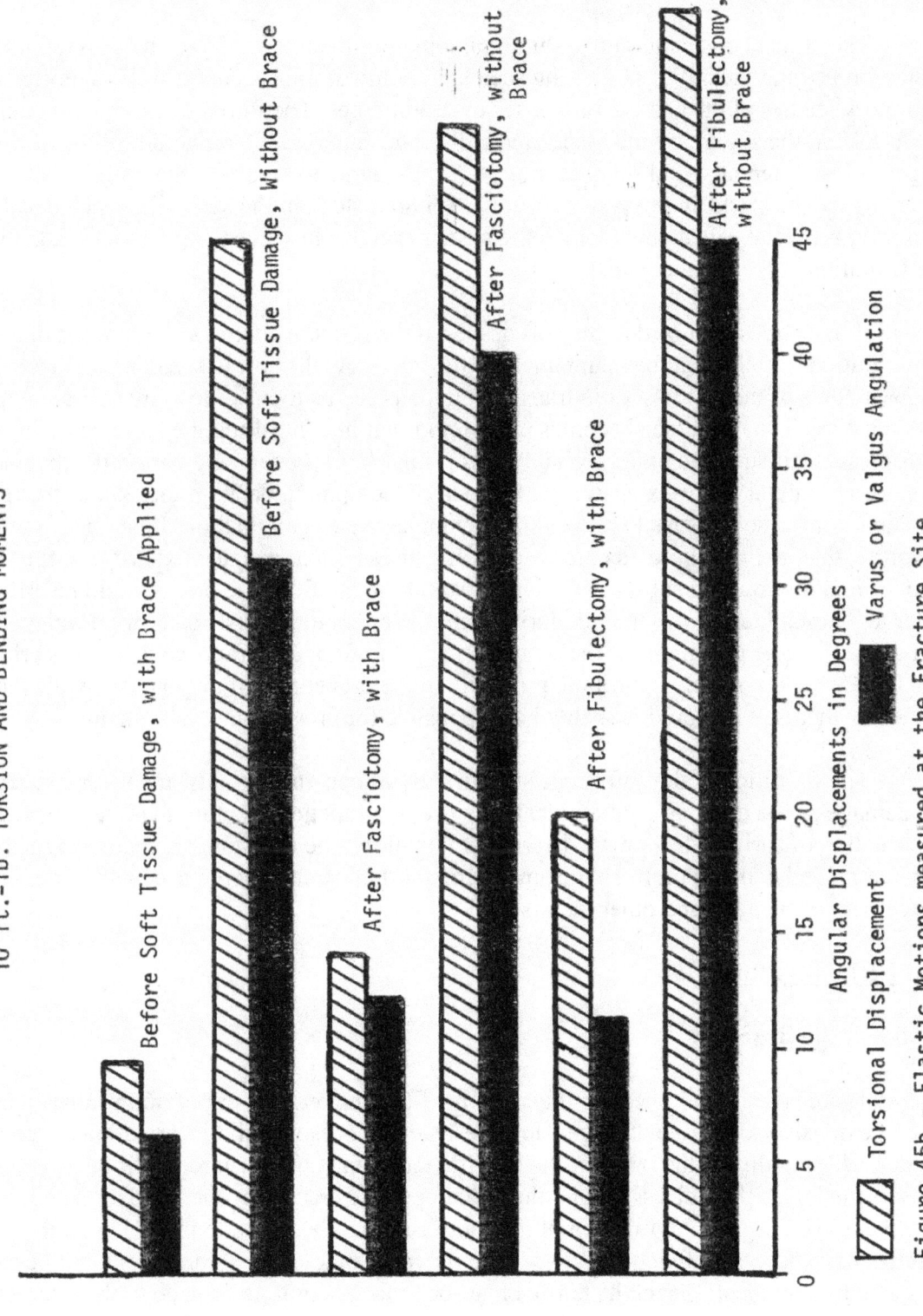

Figure 45b - Elastic Motions measured at the Fracture Site.

Conclusions and Speculations on Clinical Significance

The clinical observation that during ambulatory treatment, tibial fractures do not seem to shorten beyond the initial shortening seen at the time of injury, seems well supported by the laboratory studies of the passive properties of dead tissues. The initial displacement could cause the "slack" in the system by the associated soft tissue damage. If a reduction of the displaced fragments is at: tempted and subsequently lost, the fragments return to the initial position at injury where the slack" in the system is gone, Without slack in the system, a mechanical equilibrium is established about which the system can displace elastically under load, allowing stable motion at the fracture site.

If damage is localized in the soft tissues at the fracture site, regardless of the degree of comminution, stability can be maintained by the brace. As the soft tissue damage increases, the displacements of the soft tissues in transferring loads seems to become more hydraulic or fluid-like in nature. Thus the brace becomes more important in controlling the soft tissues in the face of extensive damage. The brace probably acts primarily to prevent angulation if soft tissue damage is minimal, but if extensive the brace is called upon to back up the skin as the rigid container for the soft tissues to prevent shortening as well as angulation. It can be assumed therefore, that in clinical practice, in cases of fractures with associated extensive soft tissue damage and/or requiring substantial length restoration, the fit of the brace would be critical in order to maintain length of the limb during ambulation, until intrinsic stability developed, If the soft tissue damage is minimal at the time of injury, the fit of the brace would be less critical. The brace would probably contribute primarily to the prevention of angulation, while the soft tissues by themselves would probably be responsible for prevention of shortening.

The condition of the soft tissue significantly affects the patients' management regime, but damage per se does not contraindicate the use of a functional below-the-knee brace and the introduction of early partial weight-bearing ambulation. The decision whether such methods are to be employed depends more on the amount of shortening considered acceptable in accordance to age, sex, occupation and other factors.

EXPERIMENT 10:

Design of the Brace

There is a strong indication that accurate fit of the brace is important, primarily in the face of extensive soft tissue damage. However, we do not have valid information concerning the desirable rigidity of the brace or the specific requirements in regard to "fit over the various parts of the leg. If the soft tissues and the cylinder of the brace prevent angulation and shortening what is then the function of the footpiece and plastic ankle joint? Do all these design features have necessary functions? The brace prevents angulatory deformities yet it bears very little bending moments during its normal load-bearing function. If damage to the interosseous membrane is minimal, the brace probably only bears minimal bending loads to prevent angulatory deformities, and the brace bears only, at most, 17% of the external axial loads applied to the limb. Hydraulic action of soft tissues primarily produces hoop stress, so how much integrity is required in the brace to accomplish these load-bearing tasks?

The following experiments were designed to study the value of each portion of the functional below-the-knee fracture-brace in -order to optimize the design of the appliance.

Procedure

Fresh above-the-knee amputation specimens were used to produce the same experimental models developed for the previous experiments. Axial compression loads, torsional moments and bending moments of similar magnitude were applied to the limbs to compare the stability of the fracture with various braces. The presently used functional below-knee brace was studied by removing or changing various parts and observing the resultant changes in stability at the fracture sight. Each procedure was reproduced in a hemi-leg model to better understand the role of the soft tissues. The leg specimen was first fitted with a conventional Orthoplast fracture brace and loads were applied. Then the proximal - one-third of the brace was removed and the loading conditions were repeated. Then the proximal two-thirds of the brace were removed and the loads repeated again. After this, all the segments of the brace were reapplied to the leg but were not fastened together so that they were allowed to move freely relative to each other. Loading conditions were then reapplied as before and instability at the fracture site noted. In the next step, all three segments of the brace were fused together and tested again. Then the footpiece was removed from the brace and the limb subjected to loading conditions Results of this test are described in Figure 46.

☐

90

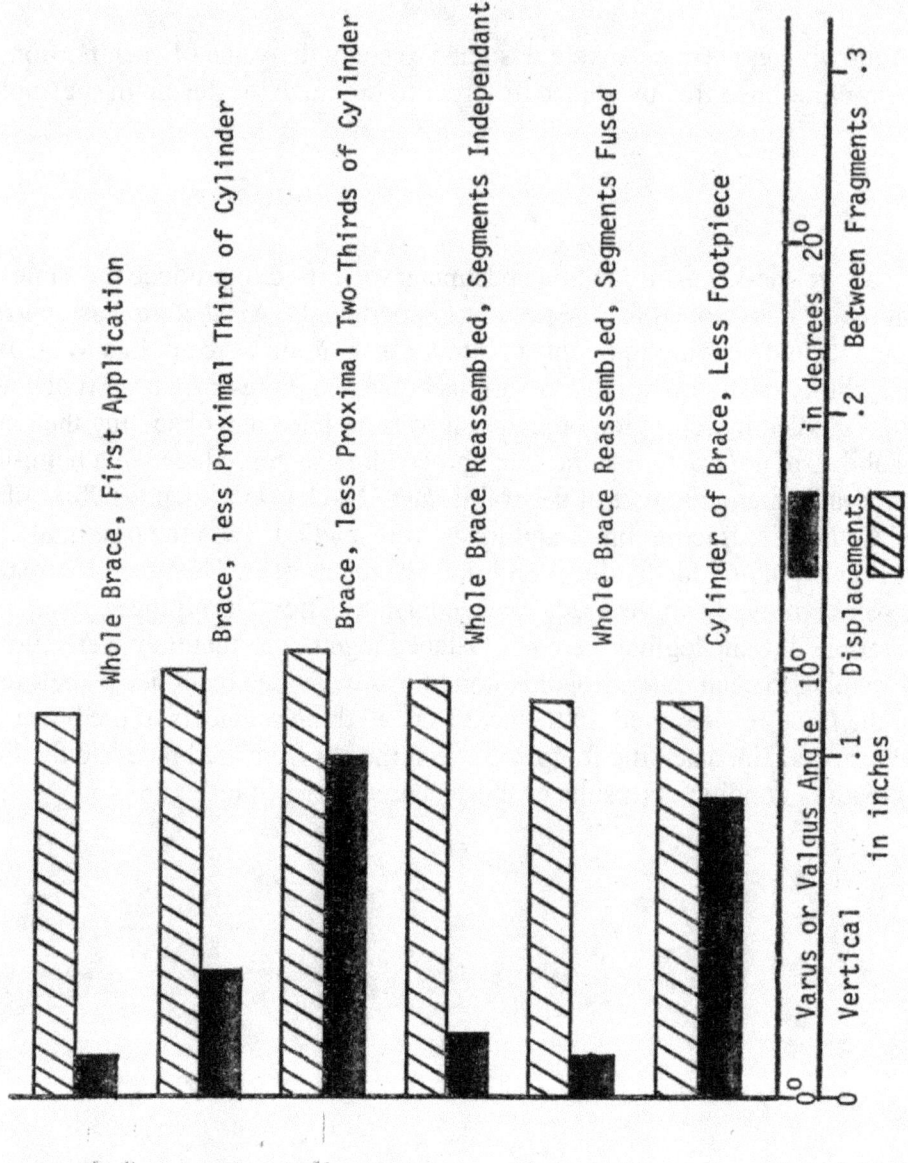

Figure 46a. Elastic Motions Measured at the Fracture Site

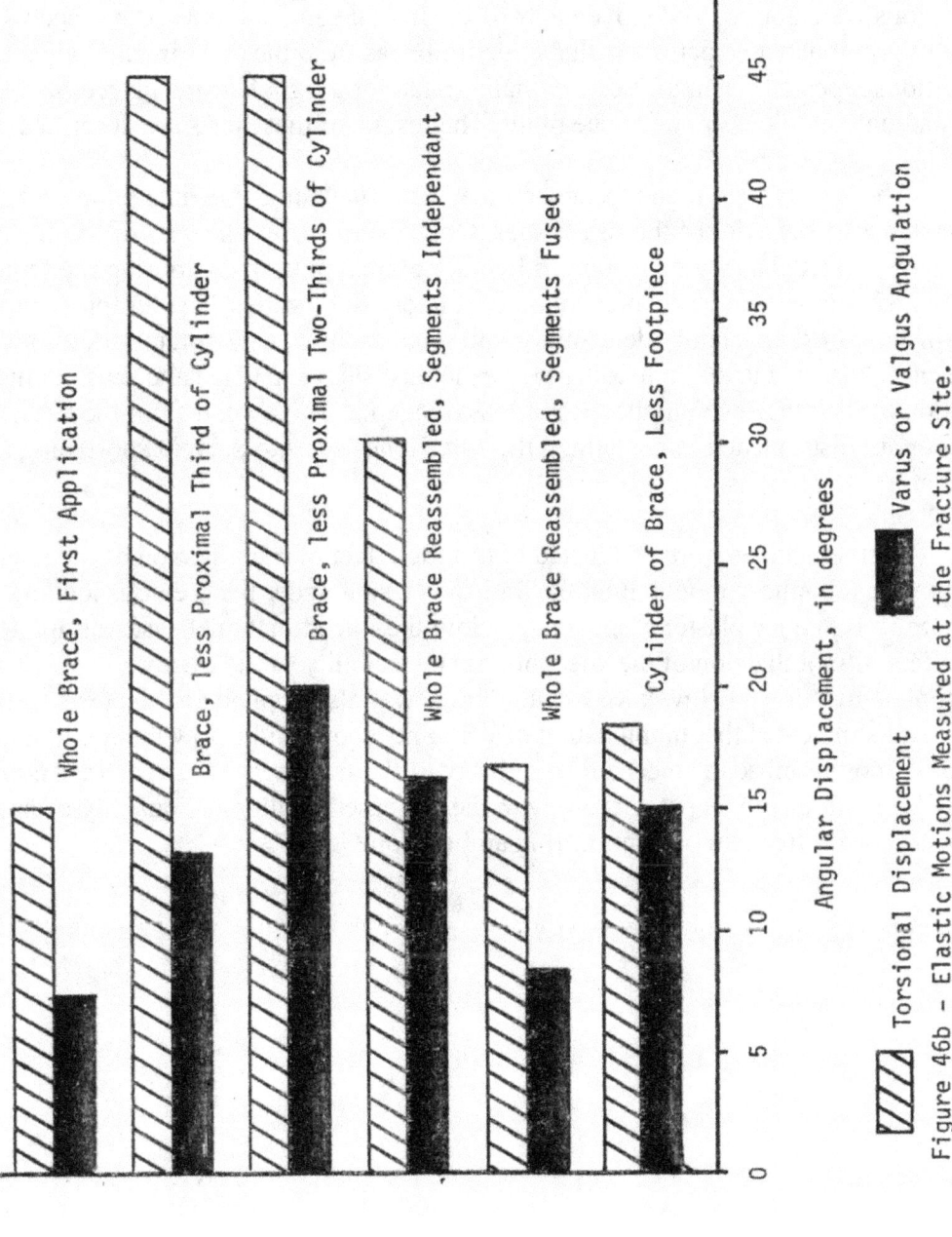

ANGULATORY AND ROTATIONAL DISPLACEMENTS DUE TO

10 ft.-1b. TORSION AND BENDING MOMENTS

Whole Brace, First Application

Brace, less Proximal Third of Cylinder

Brace, less Proximal Two-Thirds of Cylinder

Whole Brace Reassembled, Segments Independant

Whole Brace Reassembled, Segments Fused

Cylinder of Brace, Less Footpiece

Angular Displacement, in degrees

Torsional Displacement   Varus or Valgus Angulation

Figure 46b – Elastic Motions Measured at the Fracture Site.

The next experiment, tested the effect of the fit of the brace. First a limb was fractured and placed in a conventional well-fit Orthoplast below-the-knee fracture brace. Loading conditions were applied and movements of the fracture site noted. Next a brace which had been fit previously to a patient with a large limb was fit to the experimental leg and the loading conditions repeated. Then a brace originally made for a right leg was placed on a left limb and the loading conditions applied once more. The results of these tests are described in Figure 47.

These tests were then repeated on the "hemi-leg" model in an attempt to better understand the behavior of the soft tissues. Observations from the tests of the design features of the brace indicated a major role played by its proximal portion contouring the femoral condyles and the patellar tendon in the prevention of rotatory deformities. The footpiece primarily stabilizes against angulatory deformities and secondarily against rotatory deformities. Clinically, it seems to aid in holding the Orthoplast sleeve in place and thus maintains the position of the brace even in the face of loss of edema. The overall fit of the brace, its size and contour are instrumental in containing the soft tissues so as to prevent shortening and small angulatory deformities.

Further studies were conducted to test the effect of mobilization of the knee and ankle joint in our laboratory model. First the limb was tested under the previous loading conditions immobilized in a leg plaster cast. Next a below-the-knee functional cast was applied to study the effect of mobilization of the knee joint on the stability of the fracture. A conventional Orthoplast functional below knee fracture brace was also applied and the tests were repeated. The results indicated that mobilization of the joints above and below the fracture does not seriously compromise the mechanical stability of the fractures. These findings correlate well with observations in live patients 34, with the increased quality and quantity of muscle activity associated with freedom of joints during ambulation [22].

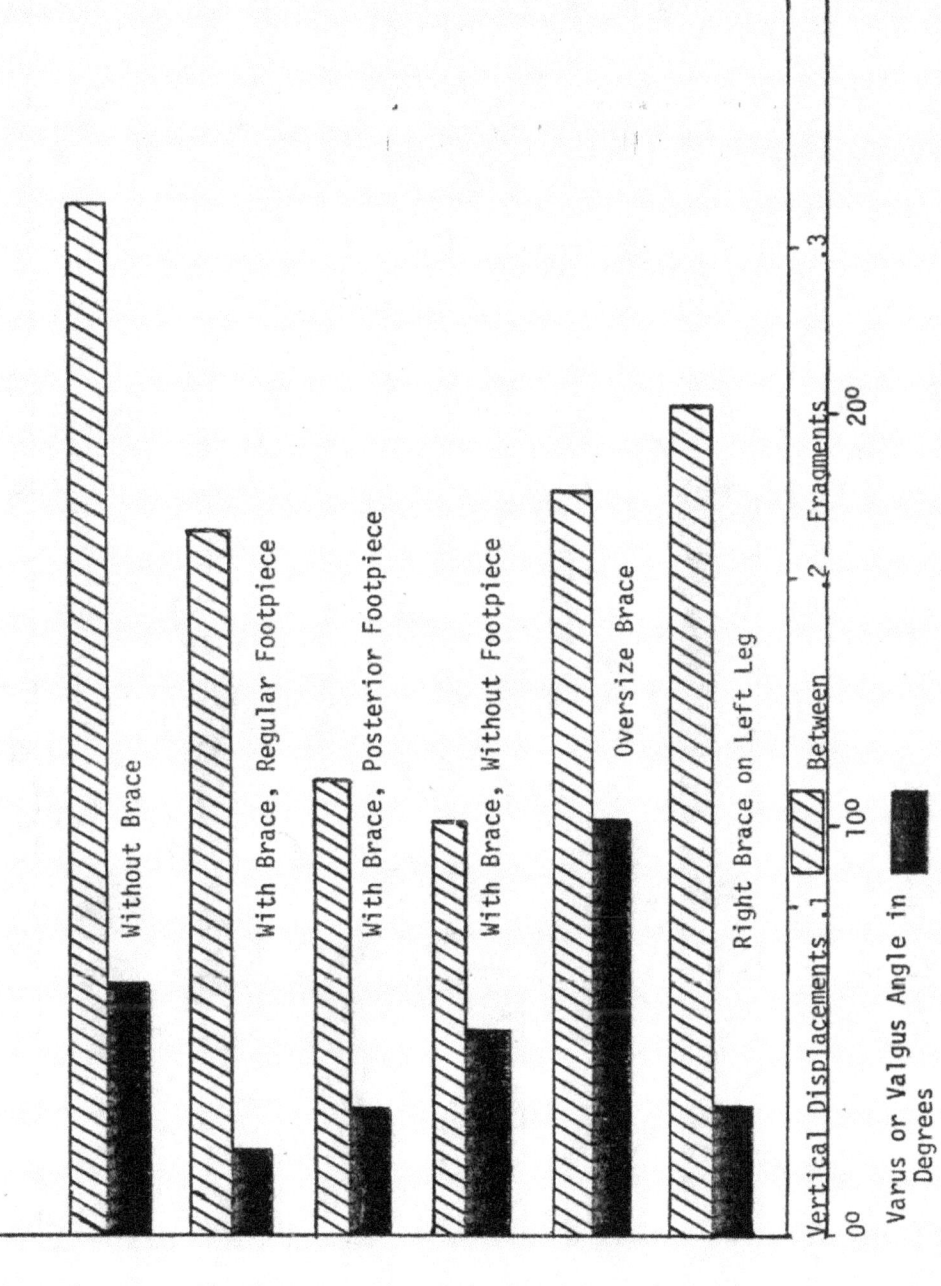

VERTICAL SHEAR AND VARUS OR VALGUS ANGULATION
FROM VERTICAL COMPRESSION LOAD OF 60 lbs.

Vertical Displacements
Varus or Valgus Angle in Degrees

Figure 47a - Elastic Motions measured at the Fracture Site.

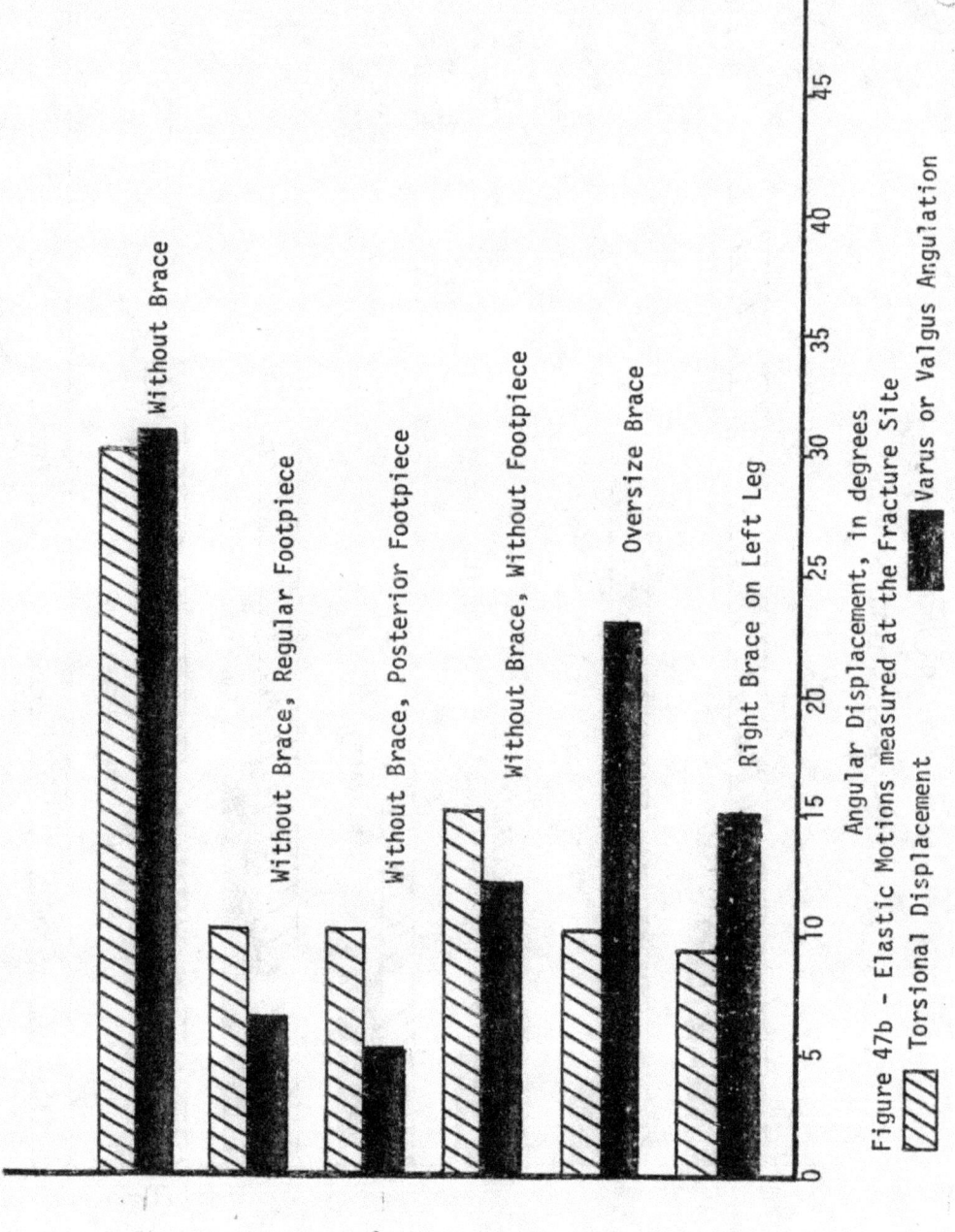

ANGULATORY AND ROTATIONAL DISPLACEMENTS DUE TO

10 ft. - 1b. TORSION AND BENDING MOMENTS

Without Brace

Without Brace, Regular Footpiece

Without Brace, Posterior Footpiece

Without Brace, Without Footpiece

Oversize Brace

Right Brace on Left Leg

Angular Displacement, in degrees

Figure 47b - Elastic Motions measured at the Fracture Site

Torsional Displacement   Varus or Valgus Angulation

The information obtained from these various tests was then used to design experimental below-the-knee fracture orthoses. Each experimental design was tested in turn on the whole-leg model under similar loading conditions. A conventional posterior leaf polypropylene ankle, foot orthoses was tested on the Orthoplast sleeve of a conventional below-the-knee fracture brace. Next the conventional ankle joint and insert for the functional fracture brace was reapplied and the results compared. Stability of the fracture was similar to either type of foot piece, but the stability was seriously compromised if no foot piece were applied to the brace. Several types of prefabricated polypropylene shells were designed based on information obtained from measurements made on over 120 different people. Those size measurements are shown in Figure 48. The first design consisted of a large anterior shell which overlapped a narrower posterior shell. The anterior shell was designed to accommodate various circumferences and contours of legs. The posterior shell was used to adjust the length of the orthoses. The next design consisted of a three-piece orthoses with three overlapping shells, one for each side of the "triangular shaped" leg. The third design used two overlapping polypropylene shells. The posterior shell encompassed most of the leg and included the proximal and distal molding of the bony prominences. The anterior shell was narrow and padded to cover the crest of the tibia. The fourth prefabricated fracture brace design consisted of two overlapping polypropylene shells with the molding of the proximal brim in the anterior shell and molding of the malleoli at the distal portion of the brace in the posterior shell (Fig. 49). The objective of this design was to obtain better length adjustment of the brace by having the proximal and distal moldings on separate portions of the brace. Each separate brace design was applied to the whole-leg model and separately tested under the same loading conditions. The results of these tests are shown in Figure 50. In general, the stability of the fractures in these limbs was not seriously affected by the "less than perfect" fit of the prefabricated braces of all designs. Therefore, it was concluded that the selection of a particular design for clinical use need be based only on the practicality of application of the various designs tested.

Ironically, regardless of the incorrect assumptions behind the original design features of the functional below-the-knee fracture cast and brace, each element of the design served a purpose and thus the latest design of fracture orthosis resembles the original quite closely.

Posterior overlapping polypropylene shells are, at this time, the most practical. Clinical evaluation of the prefabricated brace has begun but experiences are too limited to warrant their presentation. It appears, however, that it will be possible in the foreseeable future to standardize the use of prefabricated braces in the treatment of certain tibial shaft fractures. .

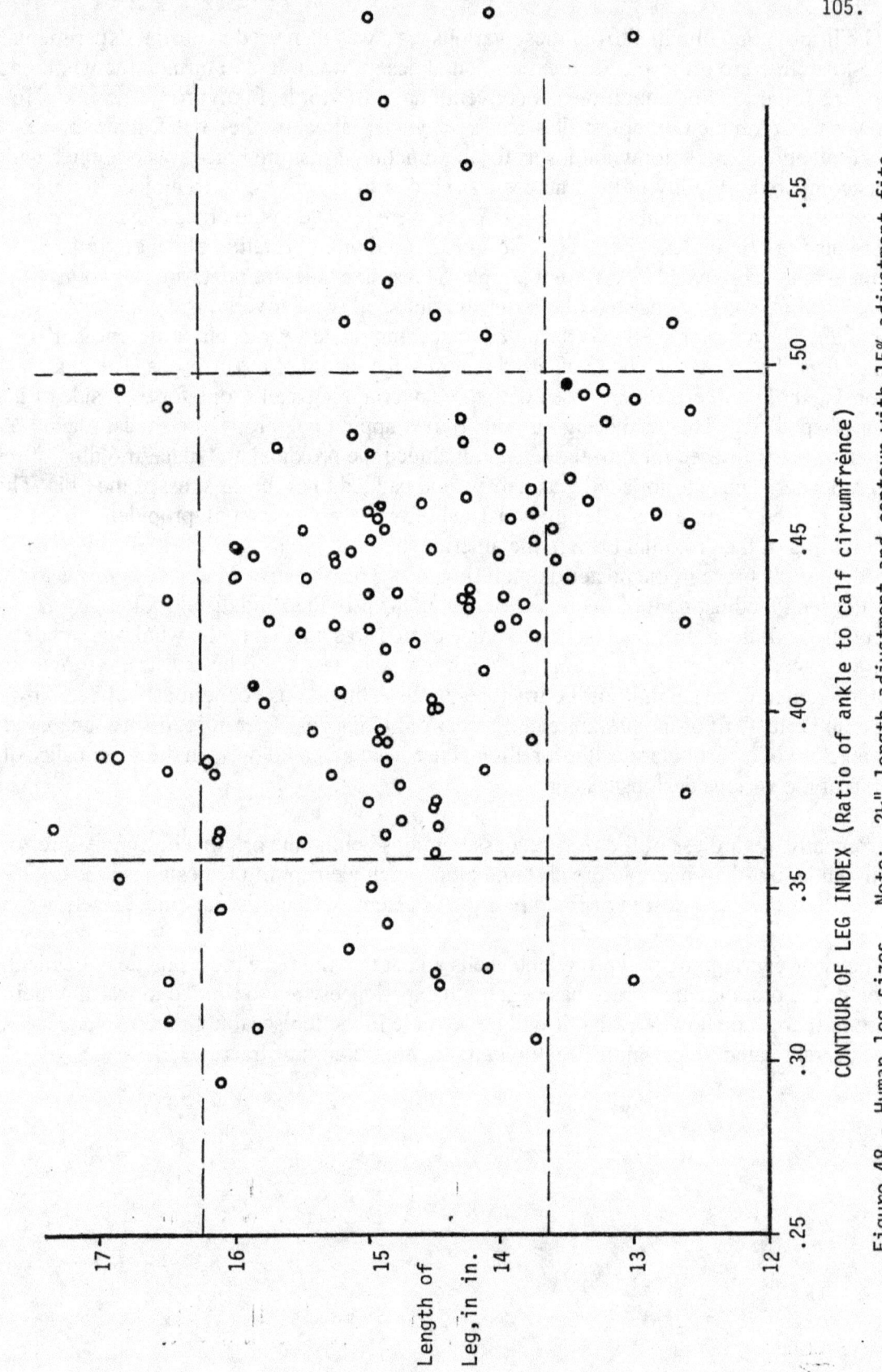

CONTOUR OF LEG INDEX (Ratio of ankle to calf circumfrence)

Figure 48 - Human leg sizes. Note: 2½" length adjustment and contour with 15% adjustment fits over 60% of patients.

From the measurements of patients and test applications of several sizes of braces to several people, it was estimated that one size prefabricated brace fits about 70-80% of the population.

It was determined that the prefabricated two-piece anterior and posterior overlapping polypropylene shells are, at this time, the most practical. Clinical evaluation of the prefabricated brace has begun but experiences are too limited to warrant their presentation. It appears, however, that it will be possible in the foreseeable future to standardize the use of prefabricated braces in the treatment of certain tibial shaft fractures.

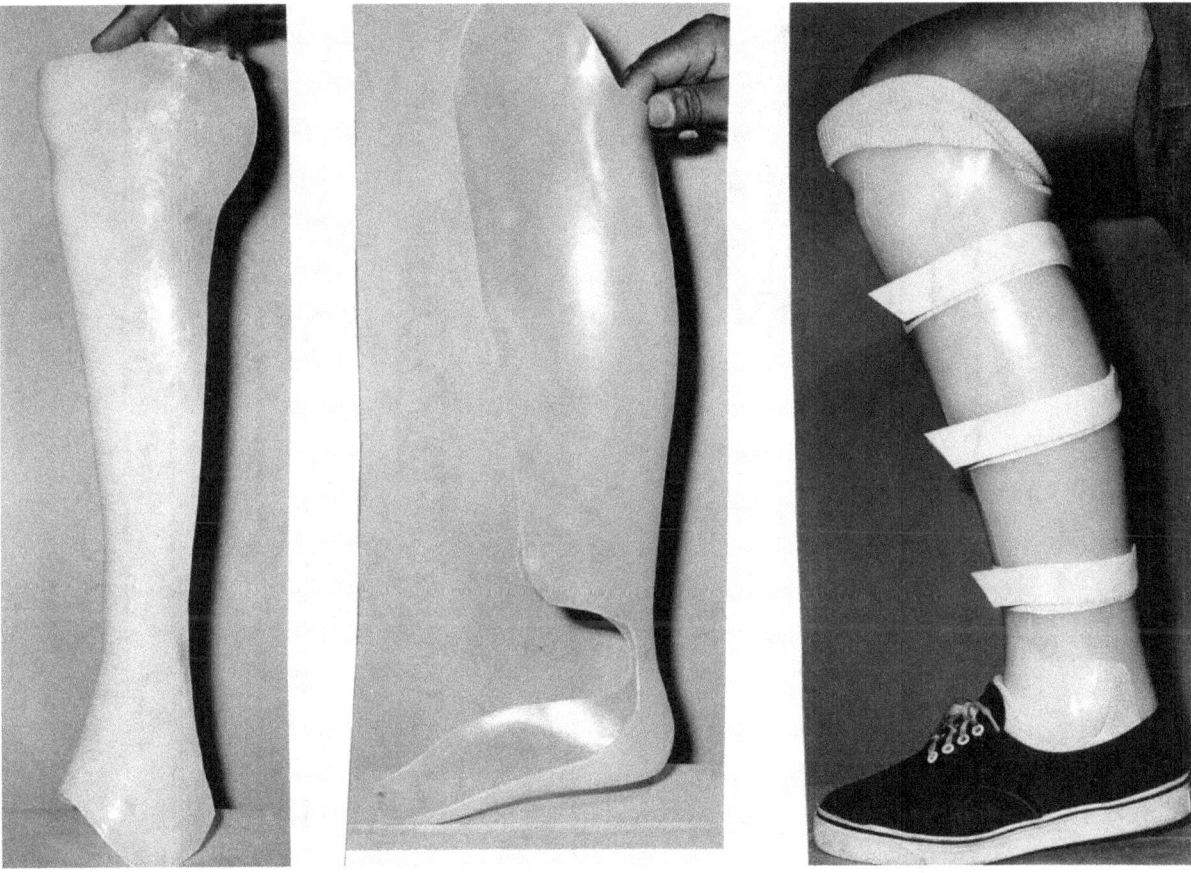

Figure 49 – Prefabricated polypropylene components for function below-the-knee fracture brace. This design is presently under clinical evaluation

VERTICAL SHEAR AND VARUS OR VALGUS ANGULATION
FROM VERTICAL COMPRESSION LOAD OF 60 lbs.

Without Brace

Orthoplast Brace

Two Piece Large Anterior Shell, PP Brace

Three Piece PP Brace

Two Piece Large Posterior Shell, PP Brace

450 lbs.

Two Piece Anterior-Posterior shell, PP Brace

Vertical Displacements .1 Between fragments .2 .3
Varus or Valgus Angle in degrees 10 20 30

99

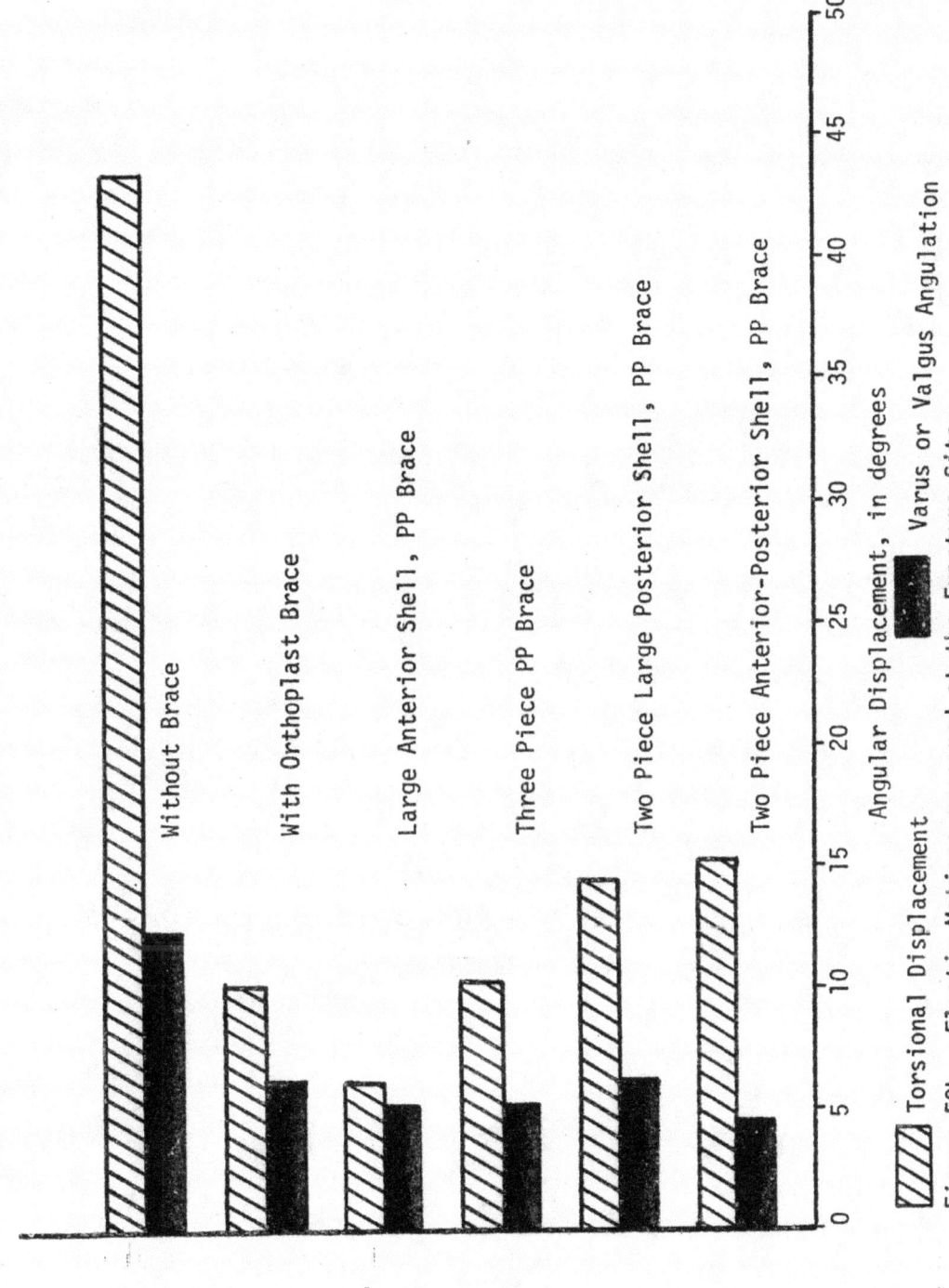

ANGULATORY AND ROTATIONAL DISPLACEMENTS DUE TO
10 ft. - 1b. TORSION AND BENDING MOMENTS

Without Brace

With Orthoplast Brace

Large Anterior Shell, PP Brace

Three Piece PP Brace

Two Piece Large Posterior Shell, PP Brace

Two Piece Anterior-Posterior Shell, PP Brace

Angular Displacement, in degrees

▨ Torsional Displacement   ■ Varus or Valgus Angulation

Figure 50b - Elastic Motions measured at the Fracture Site.

CLINICAL EXPERIENCES

Clinical Experiences

During the past eleven years we have had experiences with 532 tibial shaft fractures treated by ambulatory functional methods. Fifty-six additional patients were lost to follow-up shortly after the initiation of the treatment. The first 100 patients were treated with below-the-knee functional casts and the remaining 432 patients with below—the—knee braces. Twenty-four patients had bilateral tibial fractures. Three non-unions were encountered.

A review of the data obtained revealed that no significant difference in the results was observed between those fractures treated in the below-knee cast as compared with those treated with below-the-knee functional braces. The first 81 patients treated with below-the-knee braces had the fractures stabilized in plaster of Paris made appliances. The remaining patients were treated with Orthoplast below-the-knee functional braces.

All patients had their limbs initially stabilized in long casts extending from the base of the toes to mid-thigh with the knee joints in extension.

Only the patients who received the short, below-the-knee, functional casts or braces during the first six weeks of injury are included in this report. Therefore, this study excludes a small, though as yet undetermined number of patients, who for various reasons did not reach the fracture brace clinic during the first six weeks after the initial injury.

Ninety-one percent of the patients received their functional casts or braces during the first four weeks following the injury.

☐

We have arbitrarily established the six week period since it is our impression that the most rapid and consistent healing takes place when function is introduced during the early postoperative days. There appears to be no significant difference in the healing time of tibial fractures according to types. 'Transverse fractures healed at a median of 16 weeks (Fig. 51); comminuted fractures at 14% weeks (Figs. 52 & 53); oblique fractures at 15 weeks (Fig. 54) and segmental fractures at 17% weeks (Fig. 55). The time of healing was arbitrarily determined on the basis of radiological union, absence of pain and motion at the fracture site. The day the brace was removed coincided with the date of healing.

Only minimal difference was observed in the healing time of fractures at various levels of the tibia. Fractures of the proximal third healed at a median of 14% weeks (Fig. 56); those in the middle third at 15 weeks (Fig. 57); those on the distal third at 15 weeks (Fig. 58) and segmental fractures at 17% weeks. Of the fractures which went to union, 109 were open and healed at a median of 17% weeks (Figs. 59 & 60). Three hundred and twenty closed fractures healed at a median of 14% weeks (Fig. 61). Three hundred and eighteen tibial fractures with associated fractures of the fibula healed at a median of 15 weeks. One hundred eleven with out associated fractures of the fibula healed at a median of 16 weeks.

Nineteen percent of the patients bypassed the short-leg cast stage and went directly from the long-leg cast to the functional brace. As our experience with this method becomes greater there has been a tendency to graduate more patients directly from the long-leg cast to the functional brace. However, only fractures with minimal amount of swelling and soft tissue damage have been treated in that manner.

The maximum degree of angulation in the series was of 12° of varus; 8° of valgus and 8° of recurvatum. Ninety-two percent of the patients had less than 5% of angulatory deformity. The median shortening was 6.7 mm. Closed fractures had a median shortening of 6.4 mm and open fractures 7 mm. The maximum shortening was 2.54 cm in a patient with an open fracture and an associated fracture of the fibula. Twenty-nine patients with open fractures of the tibia and associated fibula fractures with considerable amount of shortening initially received pins above and below the fracture and had the leg immobilized in a long-leg cast. The median time of immobilization in that cast was 3 weeks.

The oldest patient was 86 and the youngest 16. Patients younger than 16 were excluded from this report.

There was one instance of thrombophlebitis, but no pulmonary emboli or anterior tibial syndromes were found.

Limitation of motion of the ankle or knee joints was not carefully recorded. However, it is our impression that permanent limitation of motion has not been encountered and that most patients regained full use of their joints and extremities prior to the removal of the brace.

40.5% of the patients sustained their fractures as passengers in either two or four heeled vehicles; 21% of the patients were pedestrians; 25% sustained their fractures as a result of falls; 11% from firearms and 2.5% from other reasons.

Two patients experienced re-fractures within the first 8 weeks following the removal of the brace. In both instances the fractures united uneventfully following reapplication of the functional brace.

Representative Examples

Figure 51 thru 61C

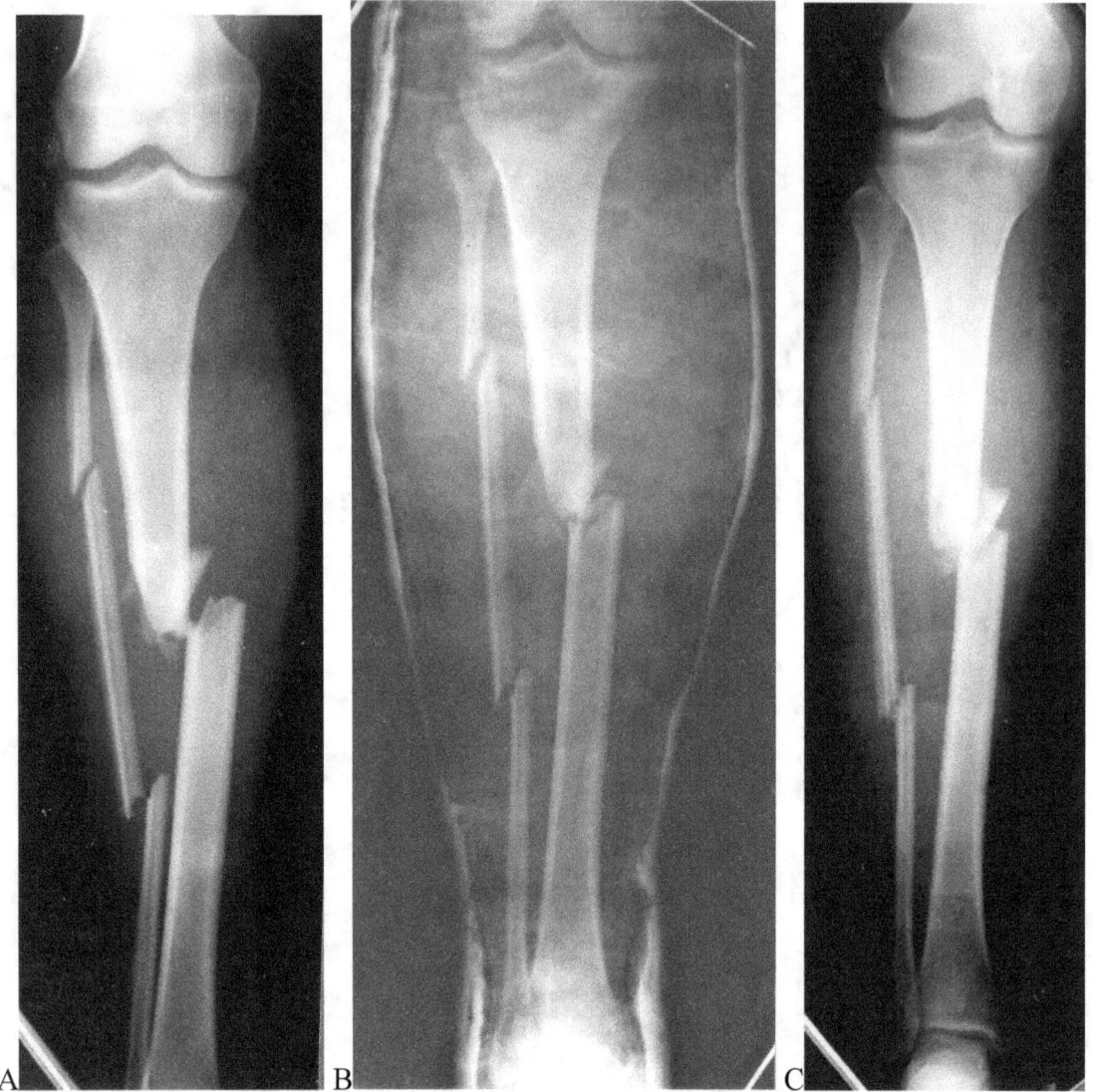

Fig 51A X-ray of transverse tibial fracture with segmental fibula after the initial accident.

Fig 51B- Appearance of the extremity following immobilization in long leg cast.

Fig 51C – X-ray through the Orthoplast brace demonstrating the alignment of the fragments .

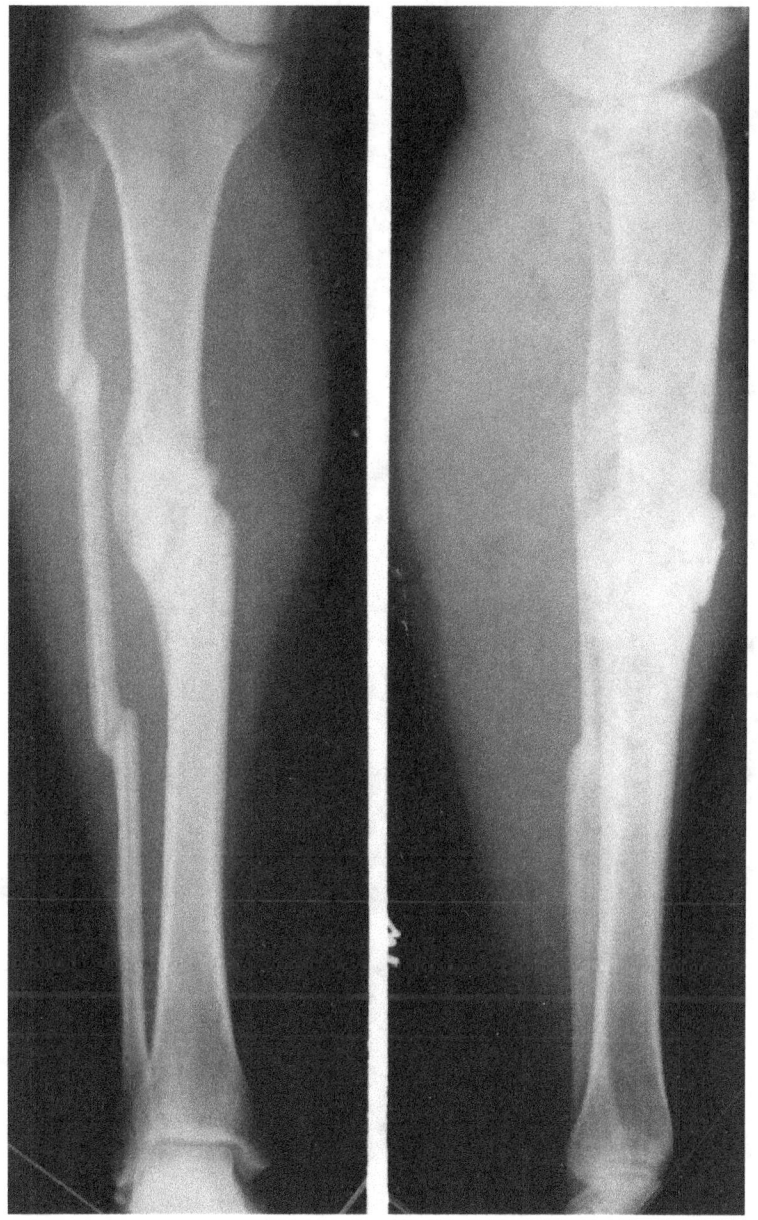

Figure 51D – Antero-posterior and lateral roentgenograms after completion of healing. Notice the maintenance of length and alignment of the tibial fragments

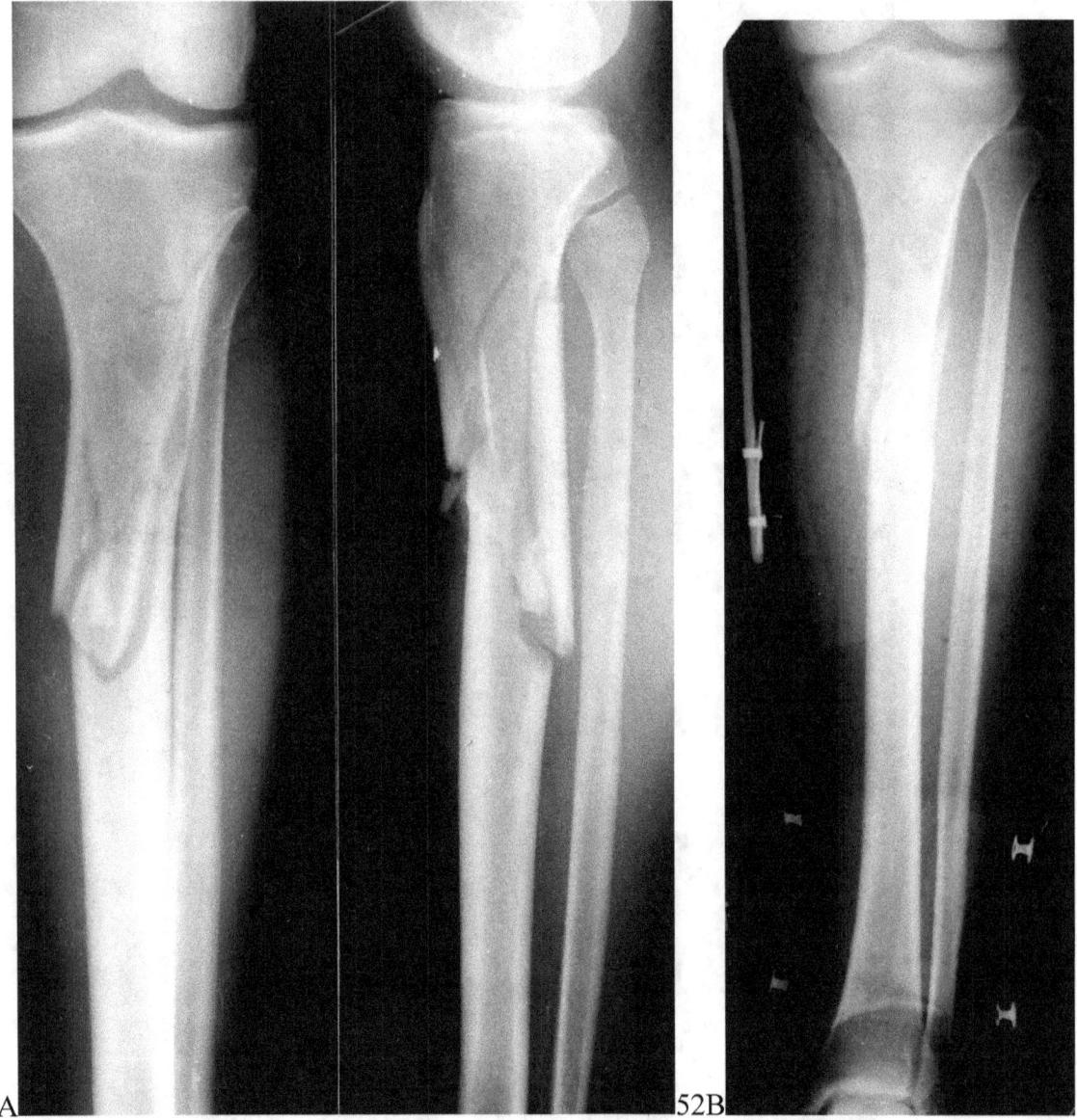

52A   52B

Fig 52A – Comminuted tibial fracture with an intact fibula.

Fig 52B – Appearance of the tibia in the functional brace

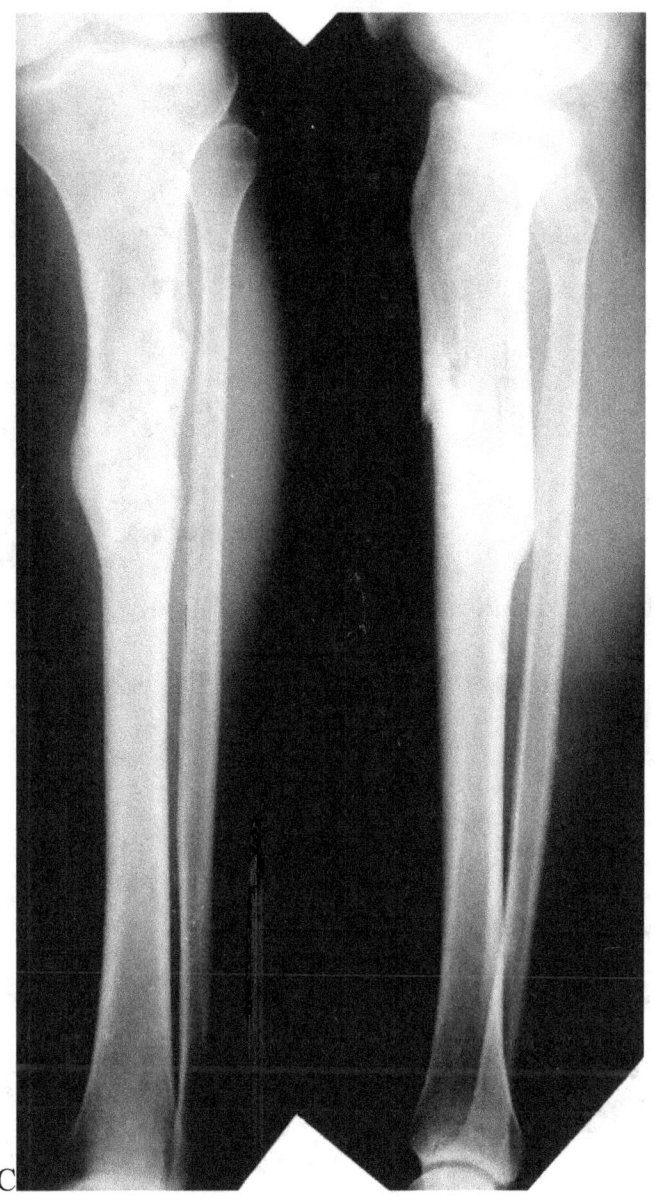

52 C

Figure 52C – Antero-posterior and lateral roentgenograms upon completion of healing

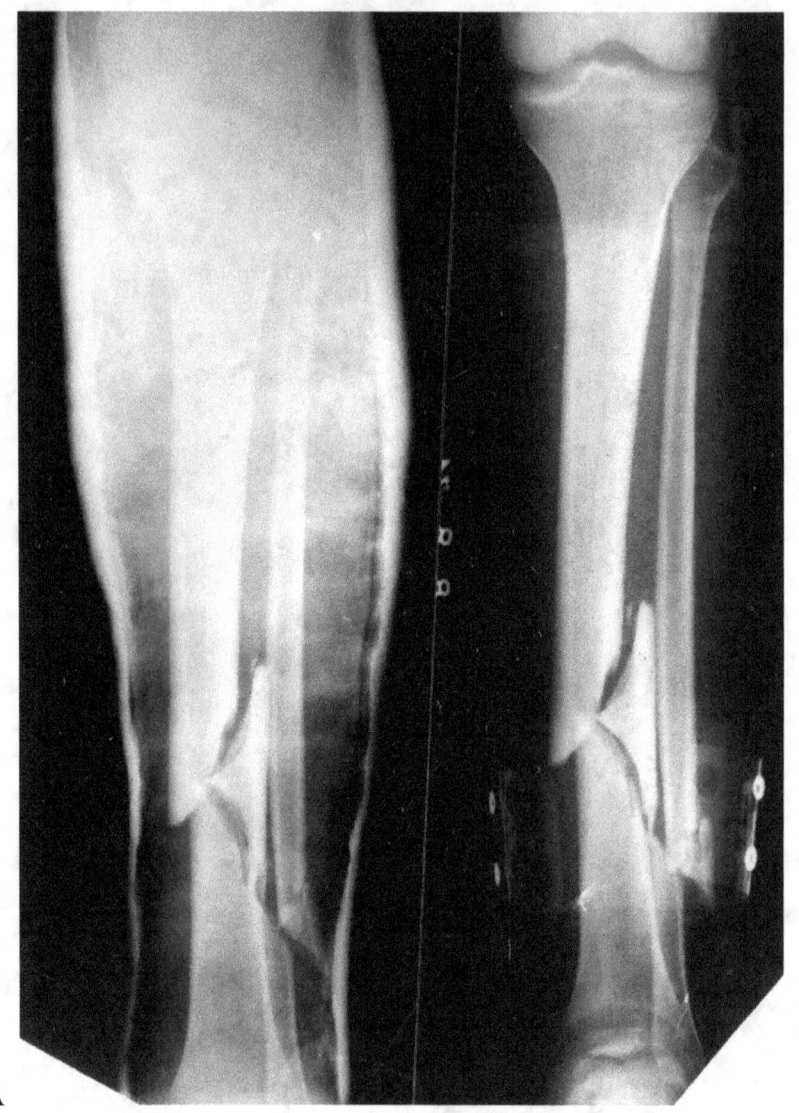

53A

FIG 53A – Comminuted fracture of the tibia and fibula

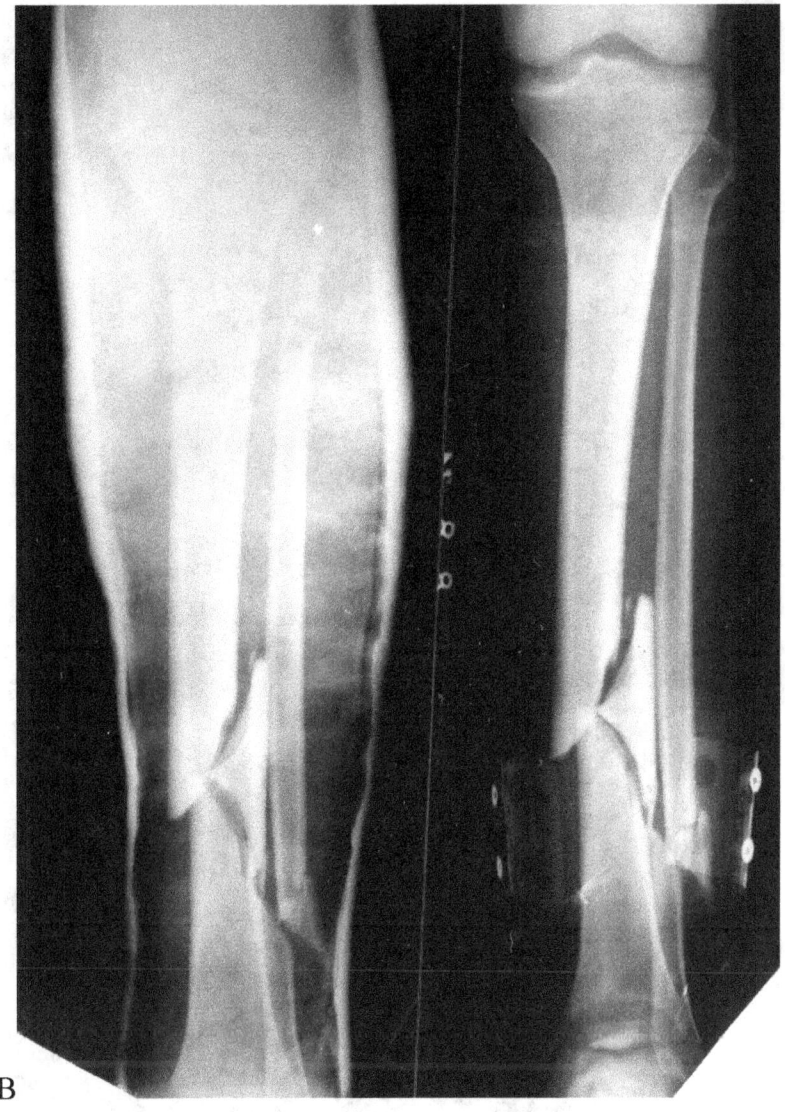

53B

Figure 53B – Roentgenograms taken through the initial plaster cast and later through the below-the-knee functional brace

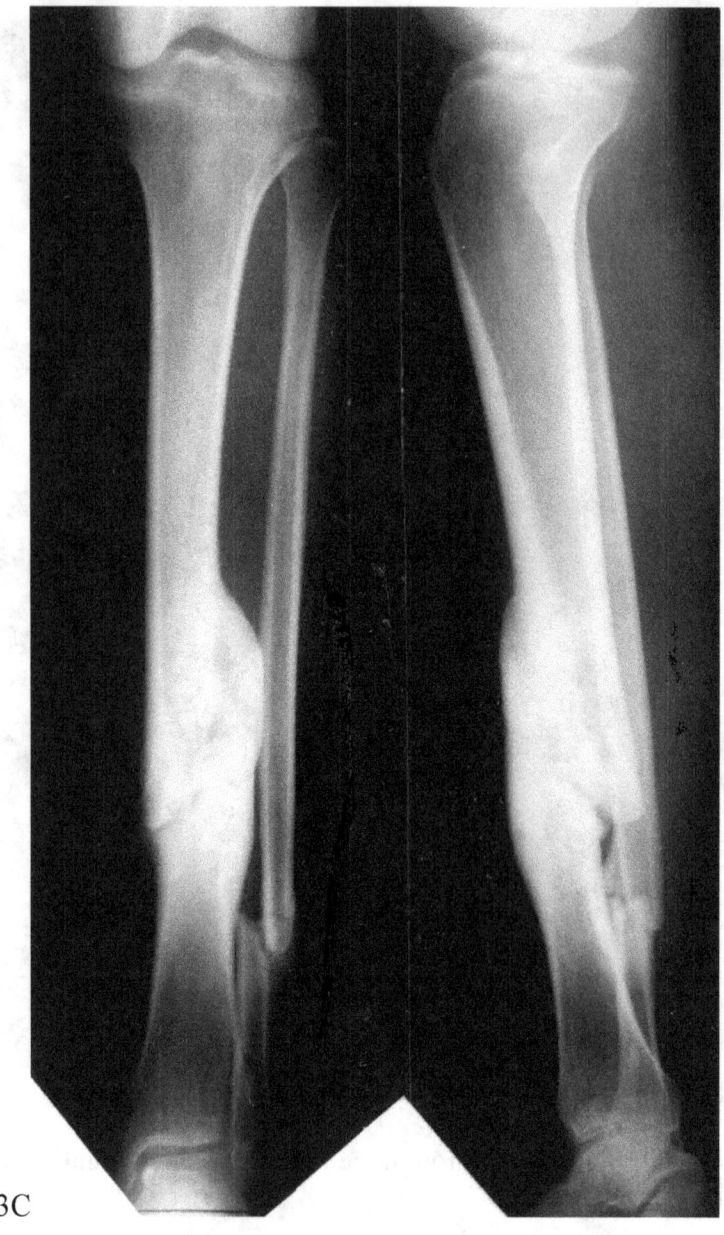

53C

Figure 53C – Antero-posterior and lateral roentgograms upon completion of healing. Notice maintenance of the length and alignment

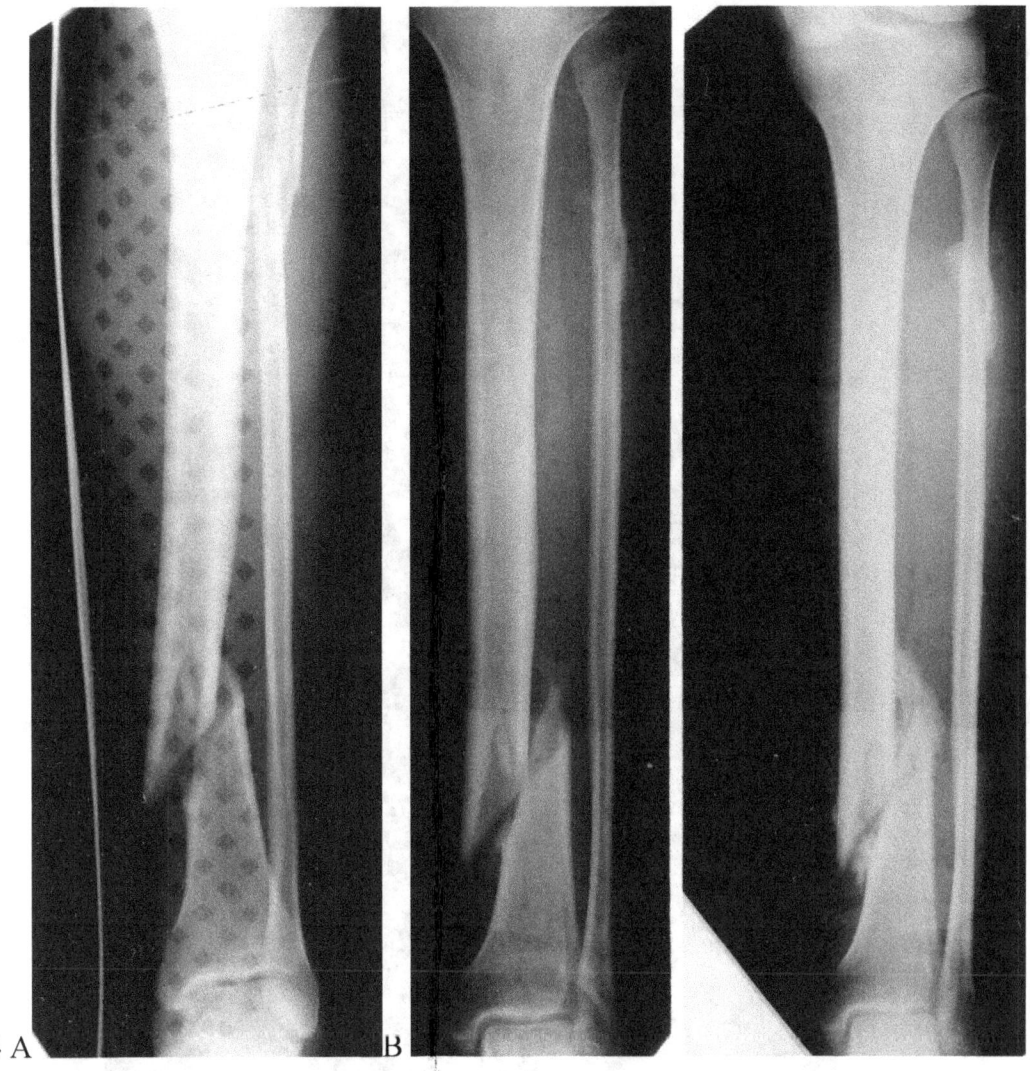

54 A   B

Figure 54 A – Oblique fracture of the distal tibia and proximal fibula upon arrival in the emergency room

Figure 54 B – X-rays obtained shortly after the application of the Orthoplast brace and later demonstrating the development of the periosteal callus

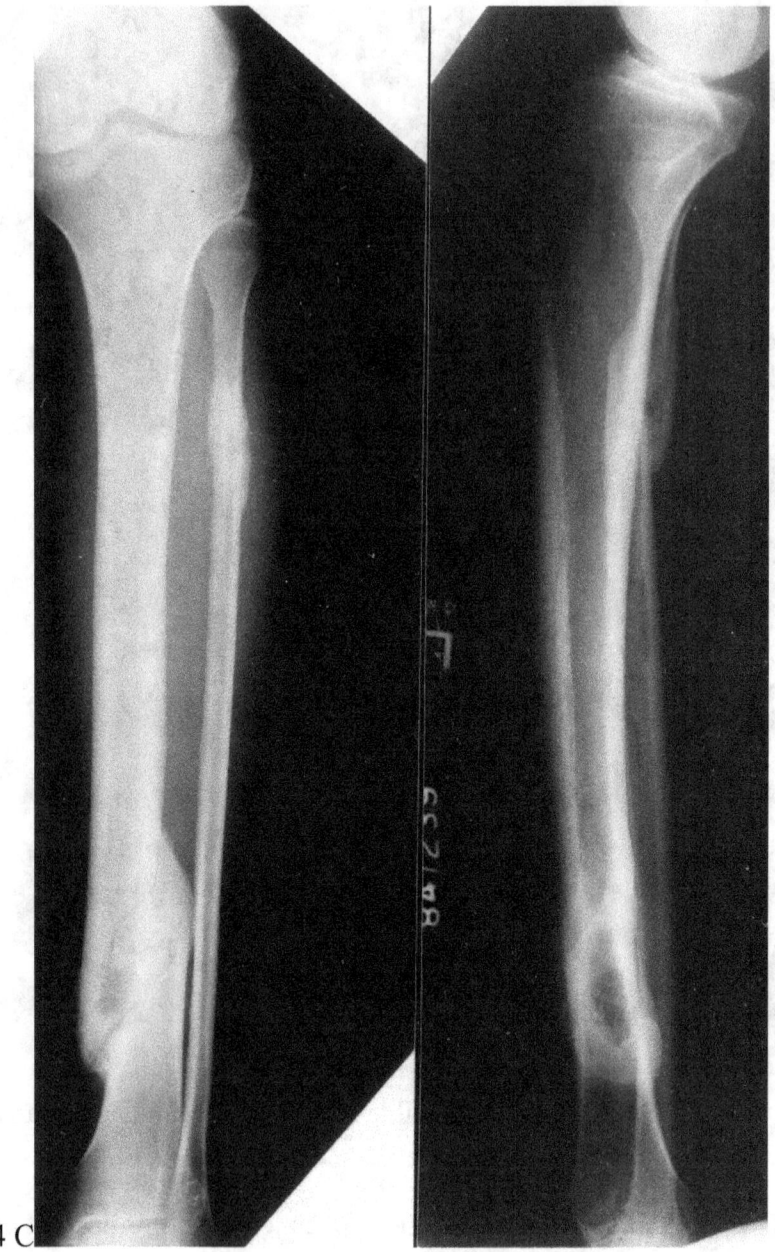

54 C

Figure 54C – Antero-posterior and lateral roentgenograms illustrating healing of the fracture without any shortening beyond that present at time of injury.

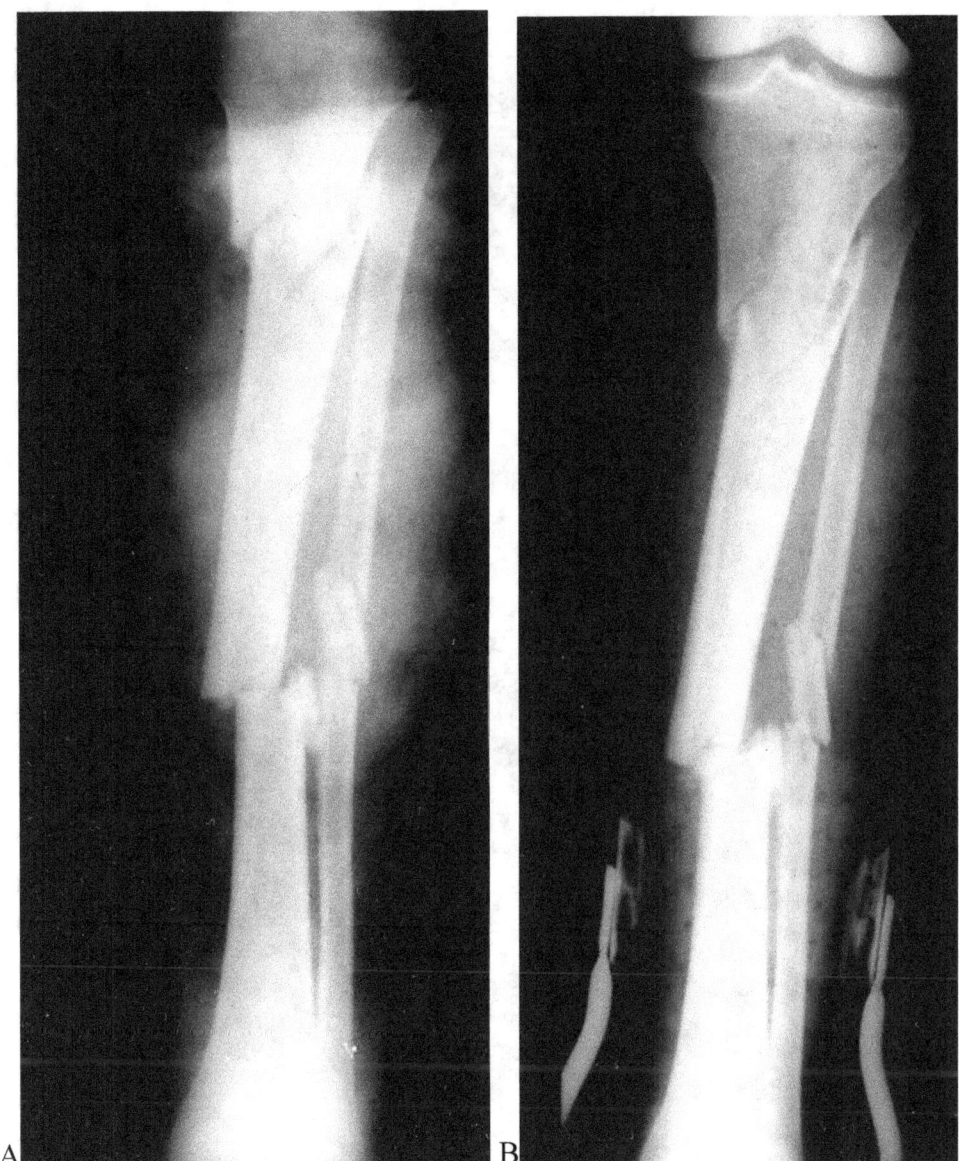

55 A   B

Figure 55A – Segmental fracture of the tibia and fibula.

Figure 55B – X-ray through the below-the-knee functional Orthoplast brace illustrating overall alignment of the fragments.

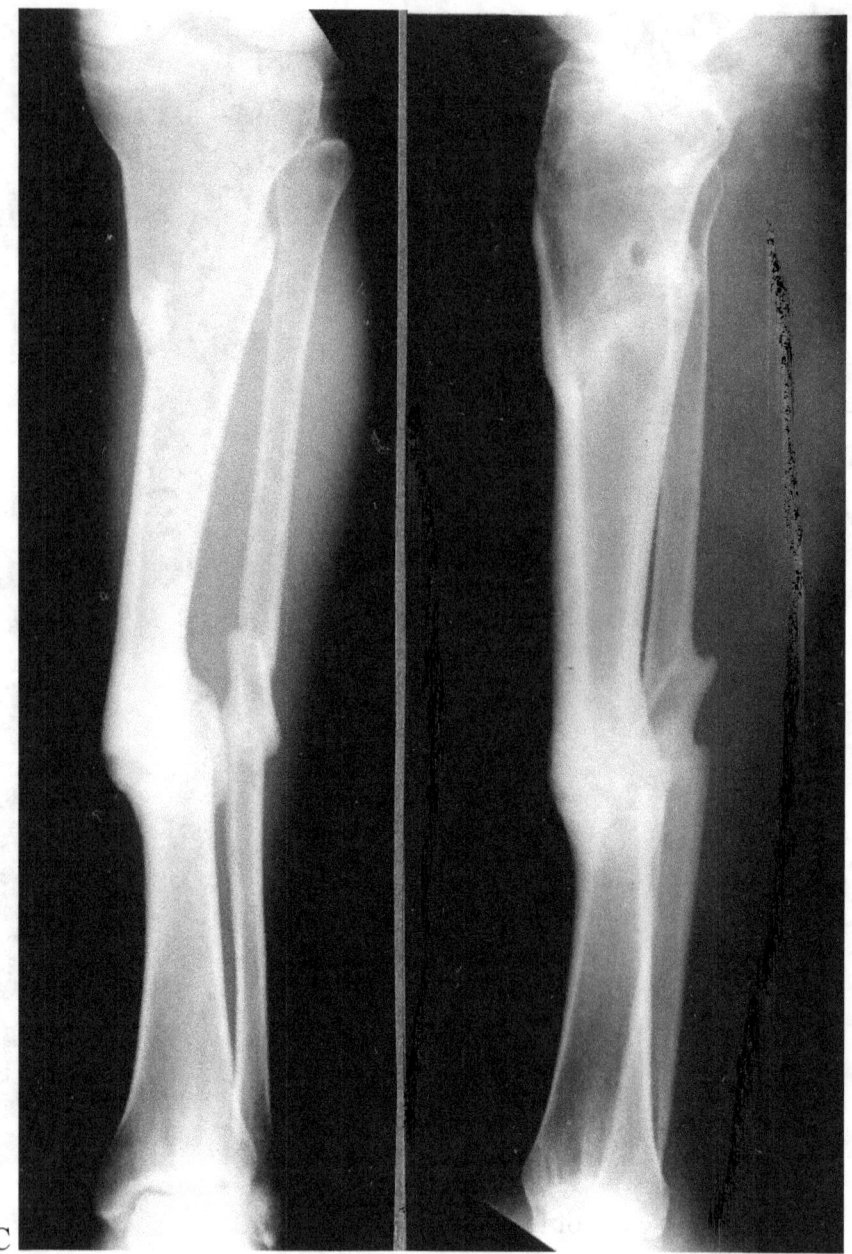

55C

Figure 55C – Antero-posterior and lateral x-rays upon completion of healing demonstrating no change in the alignment of the fragments or length of the extremity.

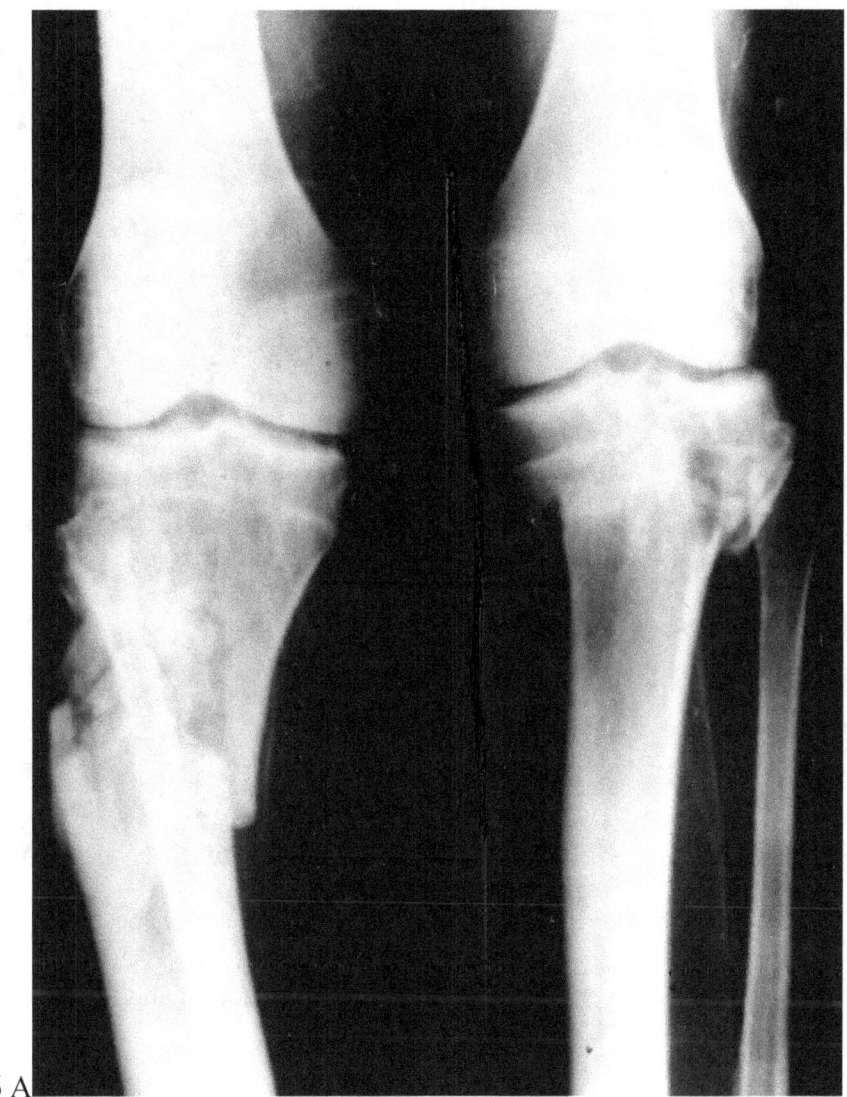

56 A

Figure 56A – Bilateral open fracture of the tibia with involvement of the joint on the left.

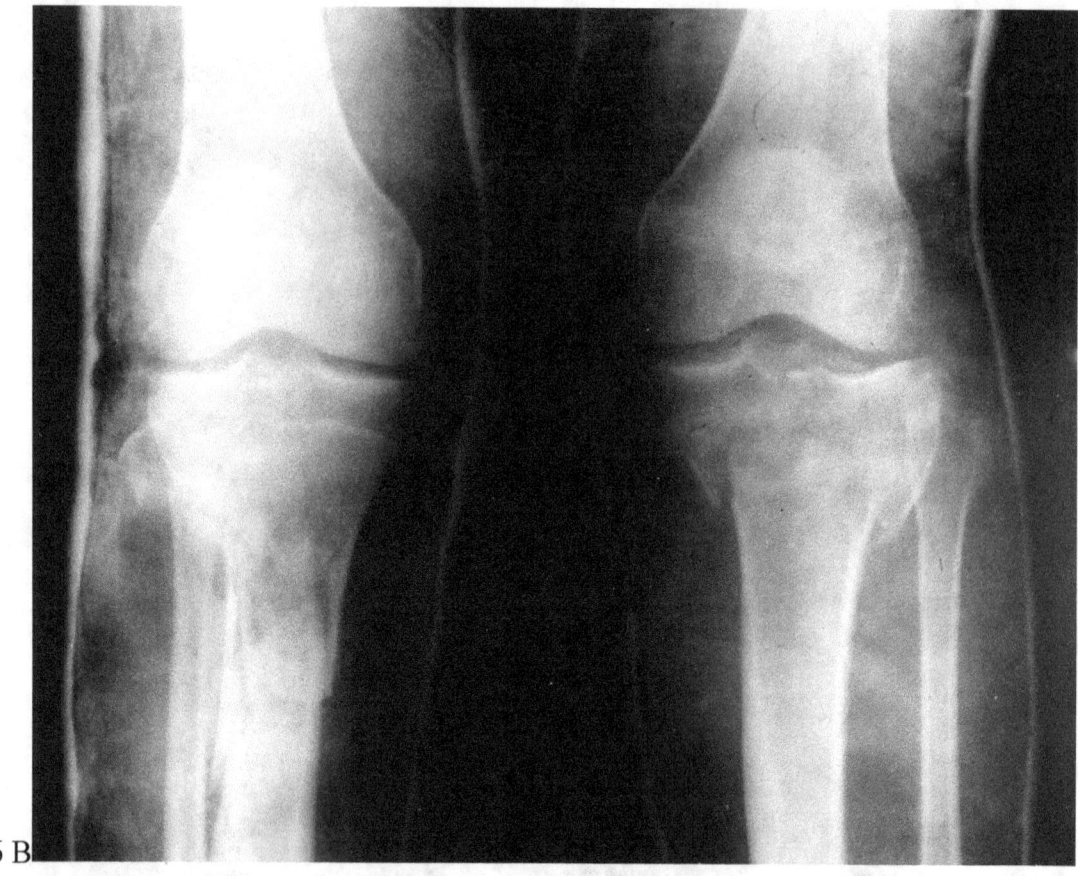

56 B

Figure 56B – Appearance of the extremities shortly after reduction and immobilization in above-the-knee casts.

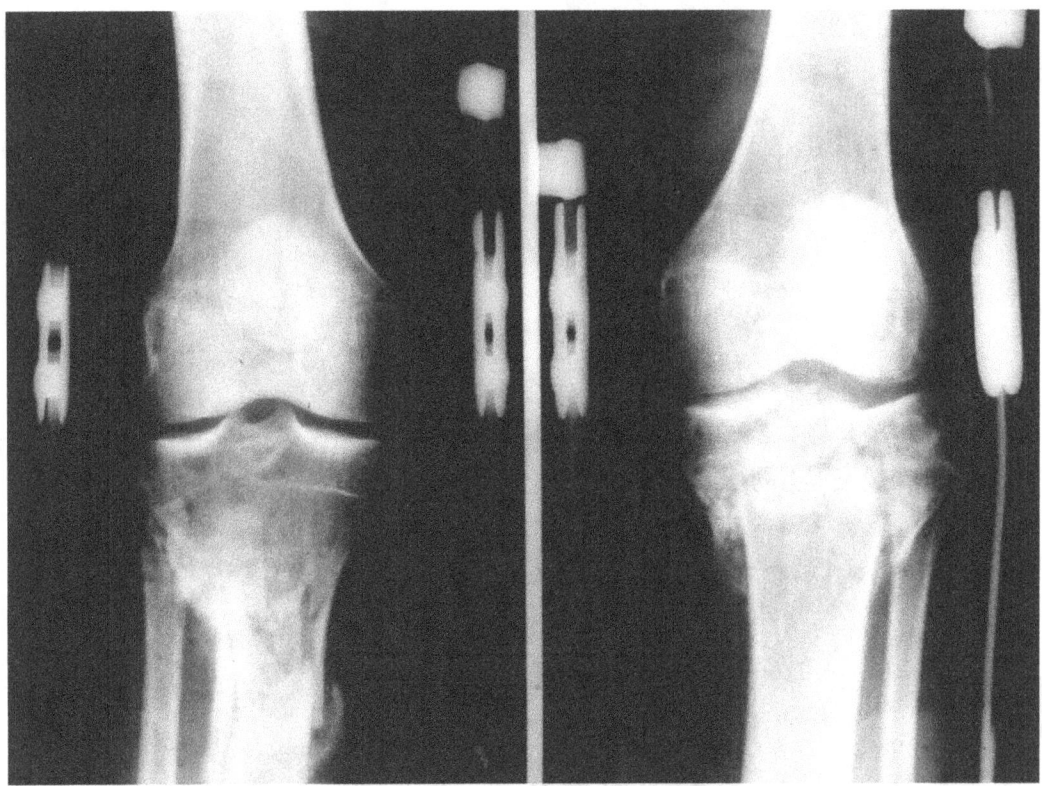

Figure 56-C Roentgenograms of both tibiae in the Orthoplast braces with thigh attachments

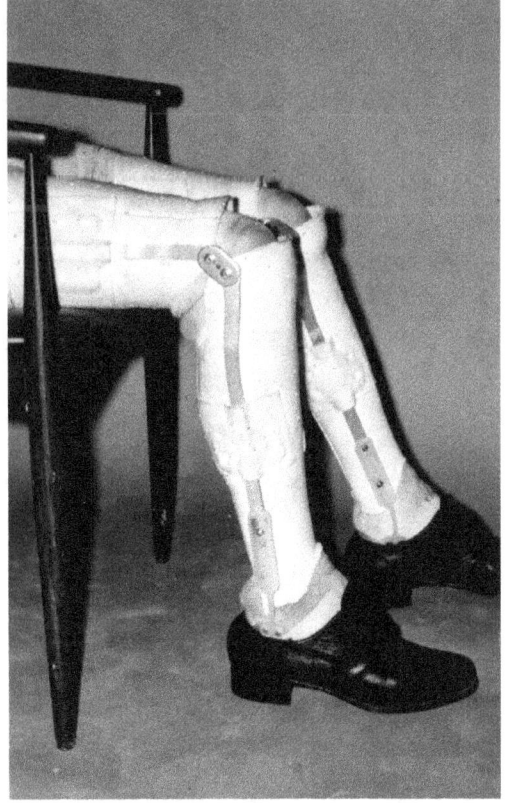

56-D – Functional braces illustrating range of motion of the knee joints

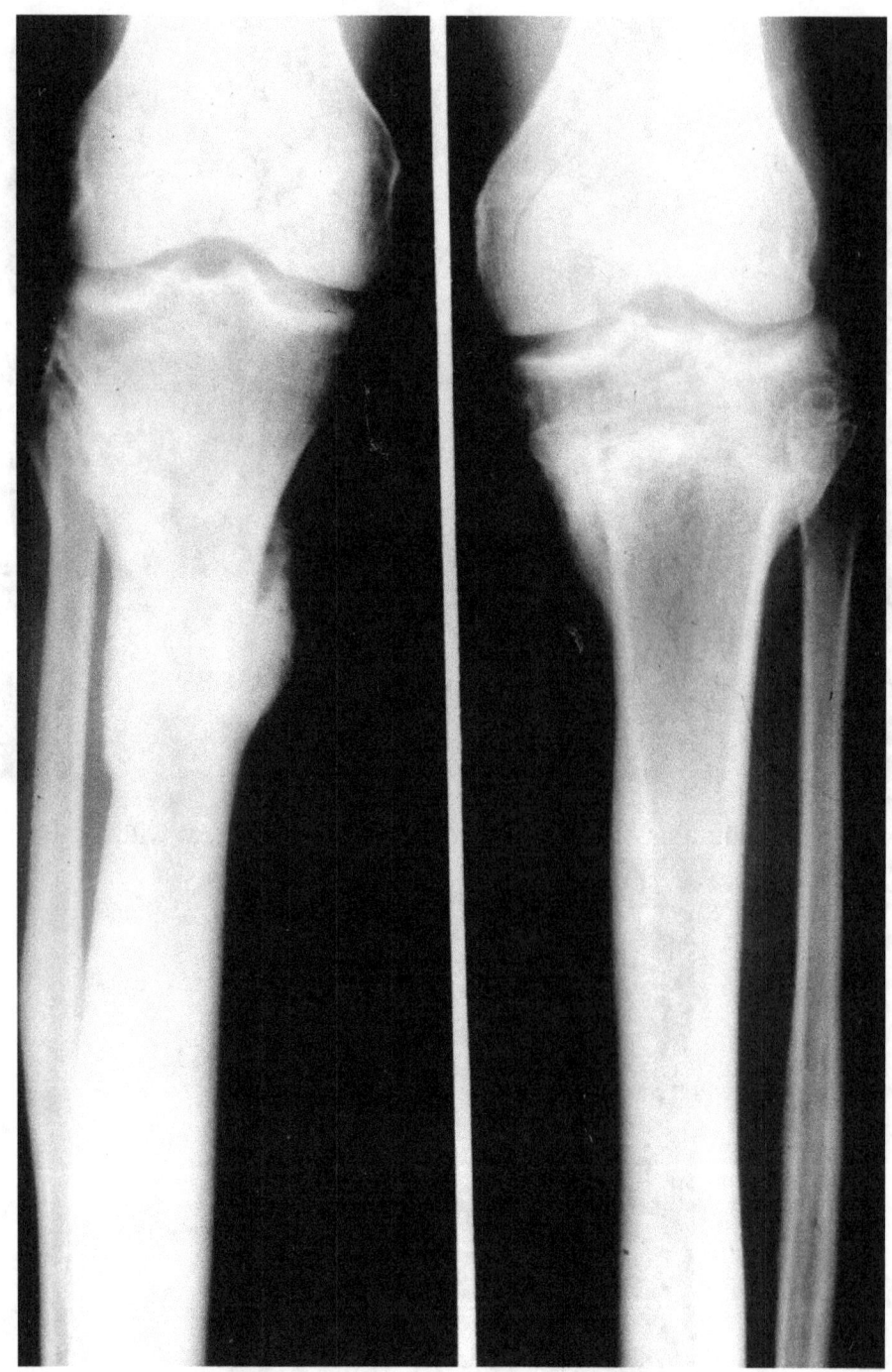

Figure 56 E – Antero-posterior roentgenograms of both tibiae upon completion of healing.

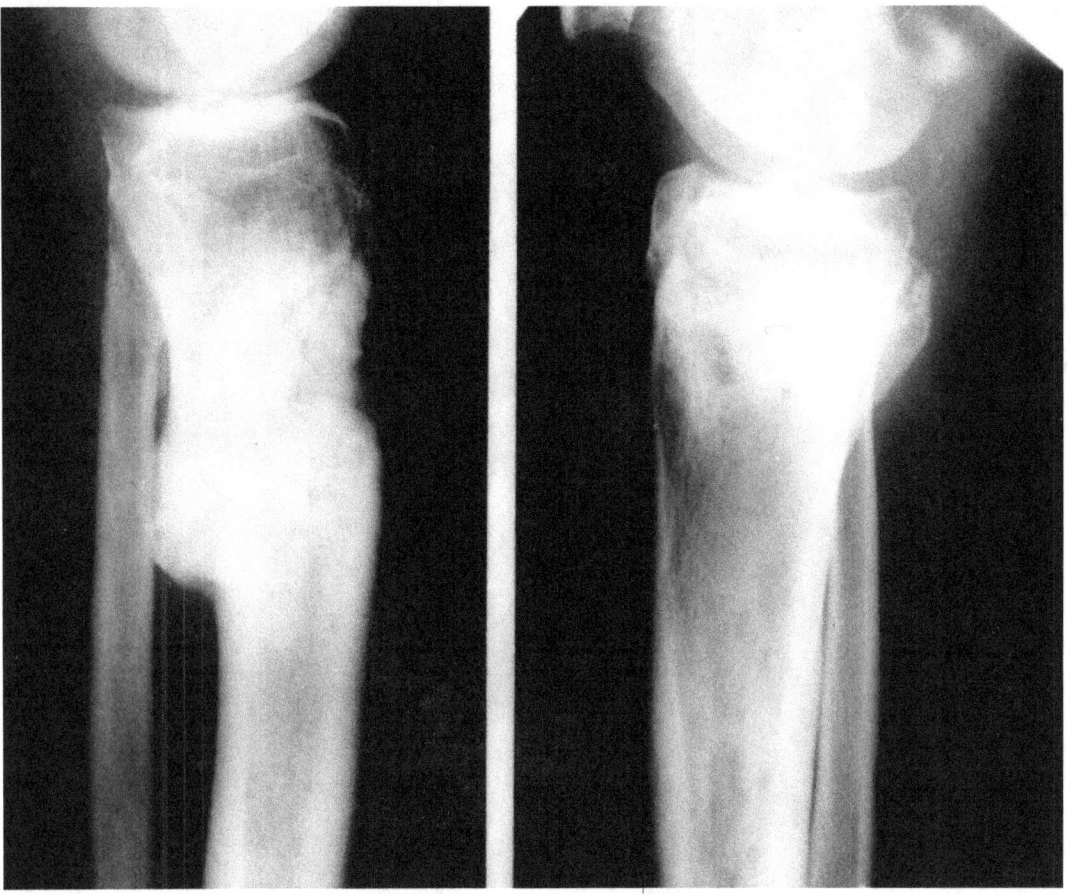

Figure 56 F – Lateral roentgenograms demonstrating the maintenance of alignment of the fragments .

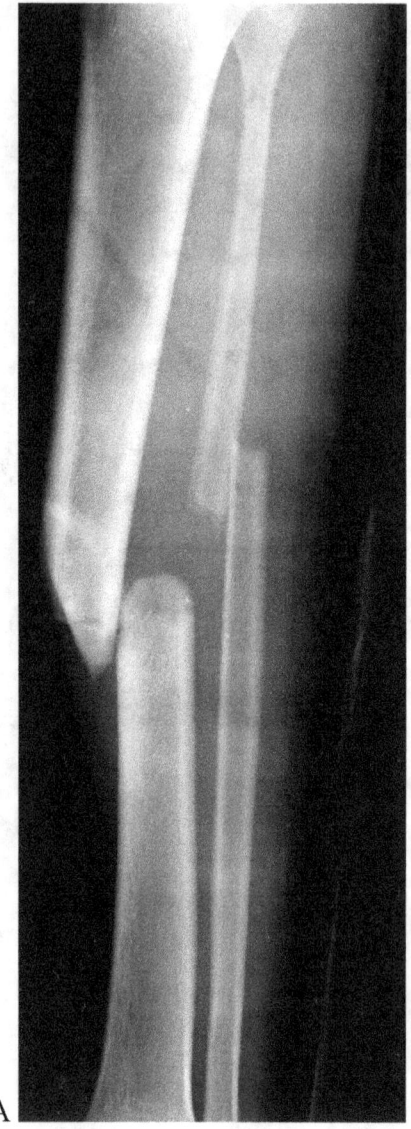

57-A

Figure 57-A Fracture of the middle third of the tibia and fibula

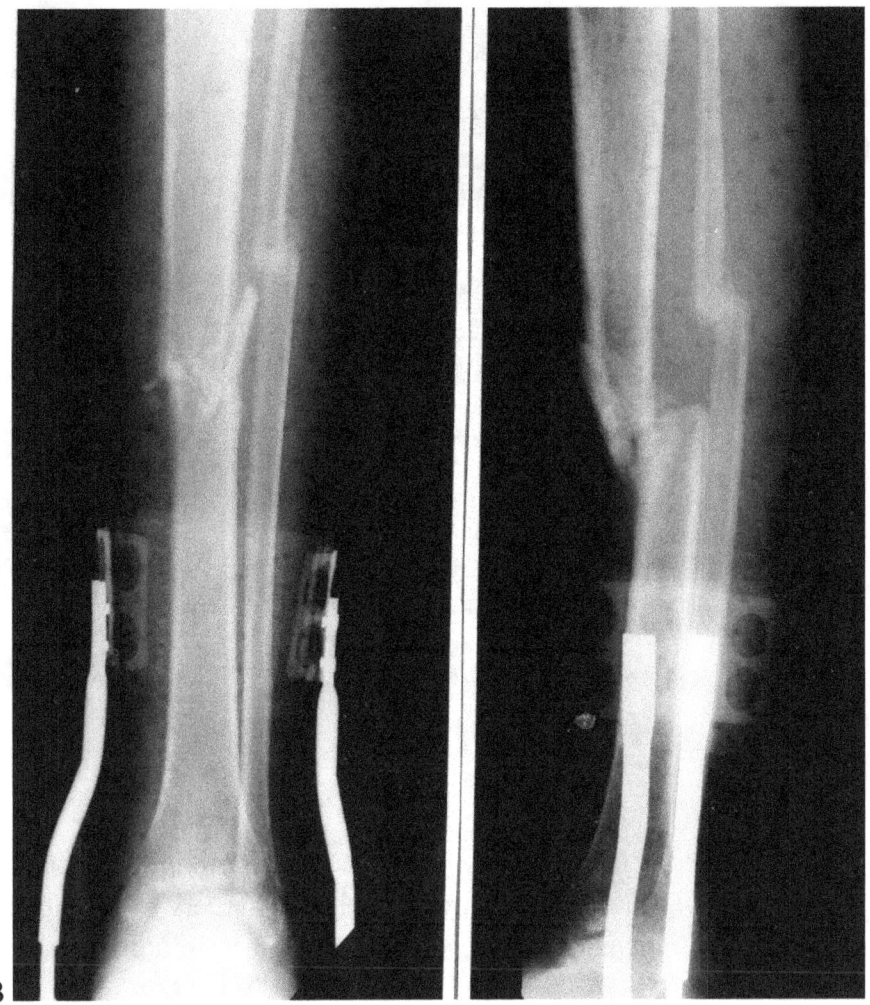

57-B

Figure 57-B – Antero-posterior and lateral roentgenograms obtained through the below-te-knee functional Orthoplast brace.

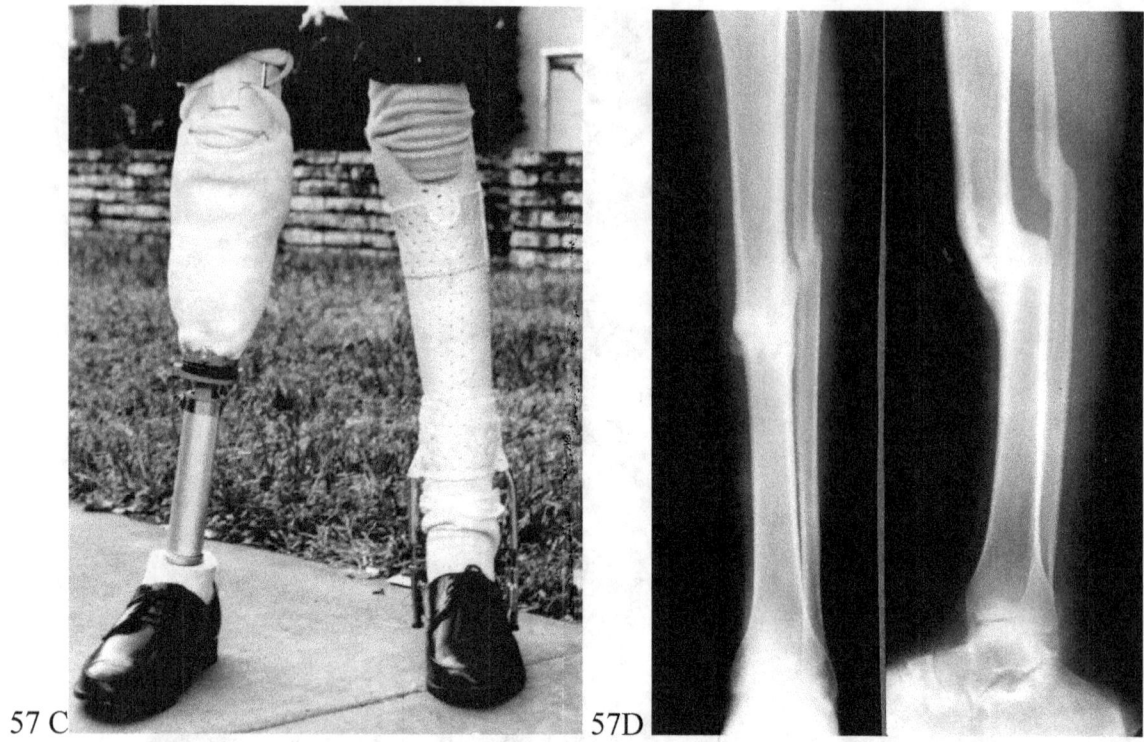

57 C   57D

Figure 57-C   Appearance of the below-the-knee functional Orthoplast brace and temporary below-the-knee amputation pylon applied in surgery following revision of the traumatic amputation sustained at the time of the initial injury.

Figure 57-D   Antero-posterior roentgenograms showing maintenance of alignment and no change in the length of the fractured tibia.

127

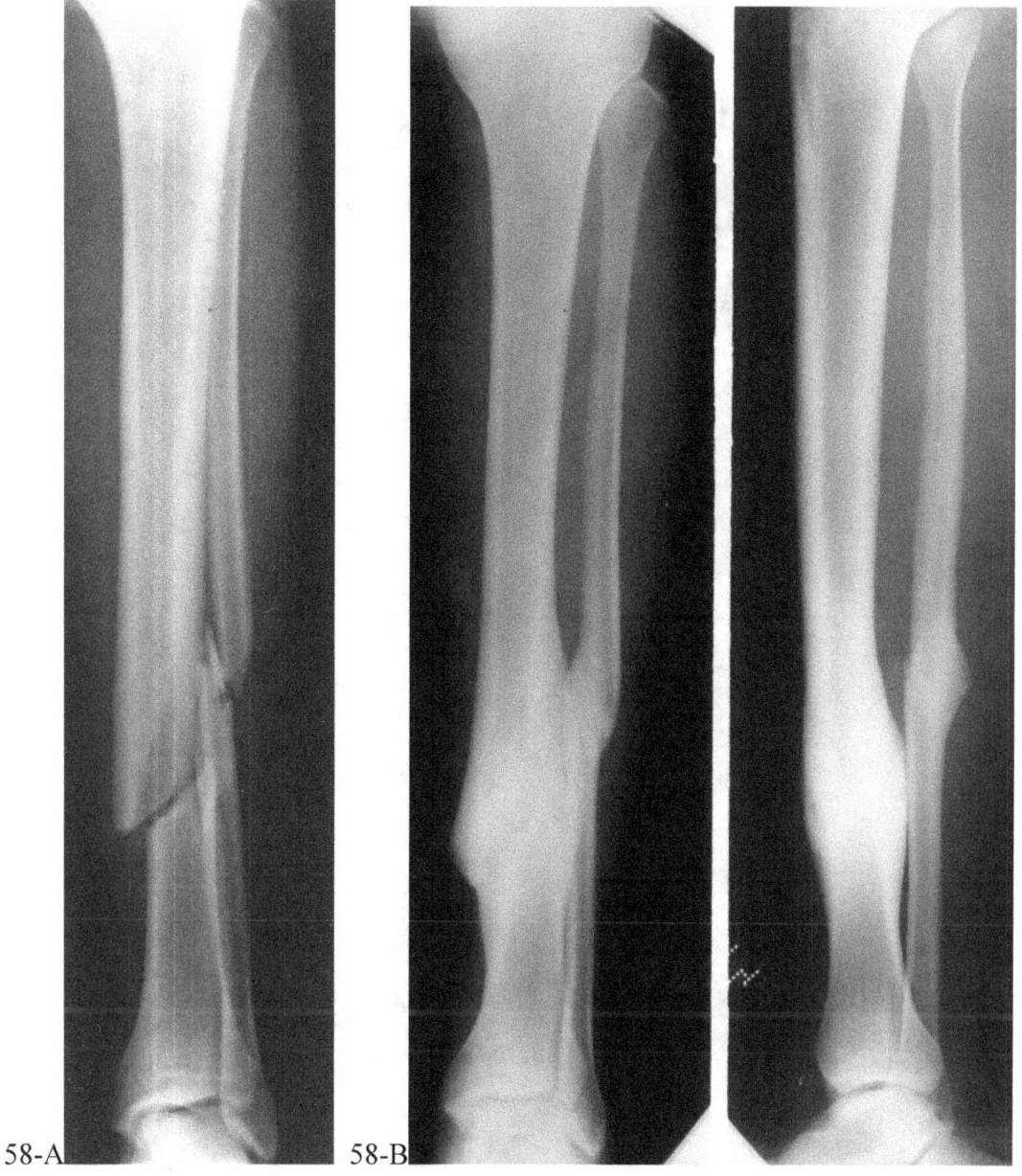

58-A   58-B

Figure 58-A   Oblique fracture of the distal third of the tibia and fibula.

Figure 58-B   Roentgenogram taken through the below-the-knee functional Orthoplast brace.

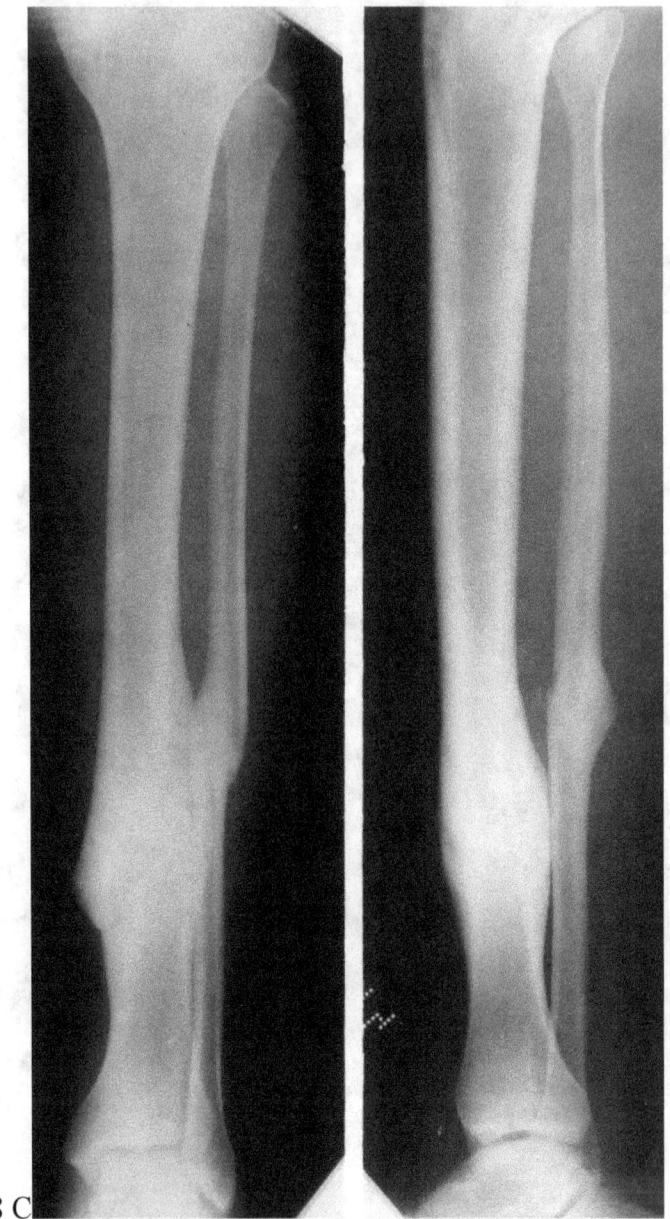

58 C

Figure 58-C   Antero-posterior and lateral roentgenograms upon completion of healing
demonstrating maintenance of length and alignment of the extremity.

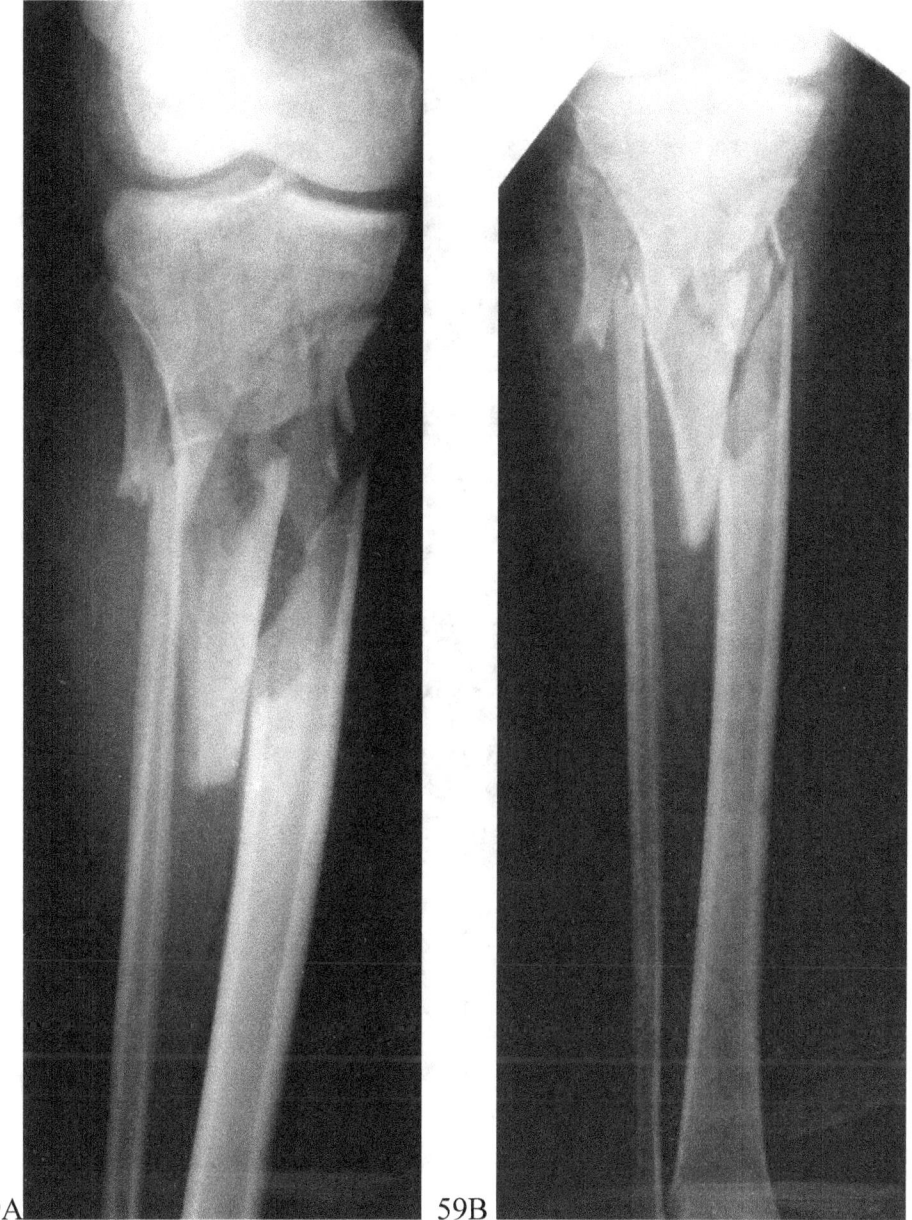

59A   59B

Figure 59-A   Open comminuted fracture of the proximal third of the tibia and fibula.

Figure 59-B   Roentgenogram obtained through the below-the-knee functional Orthoplast brace.

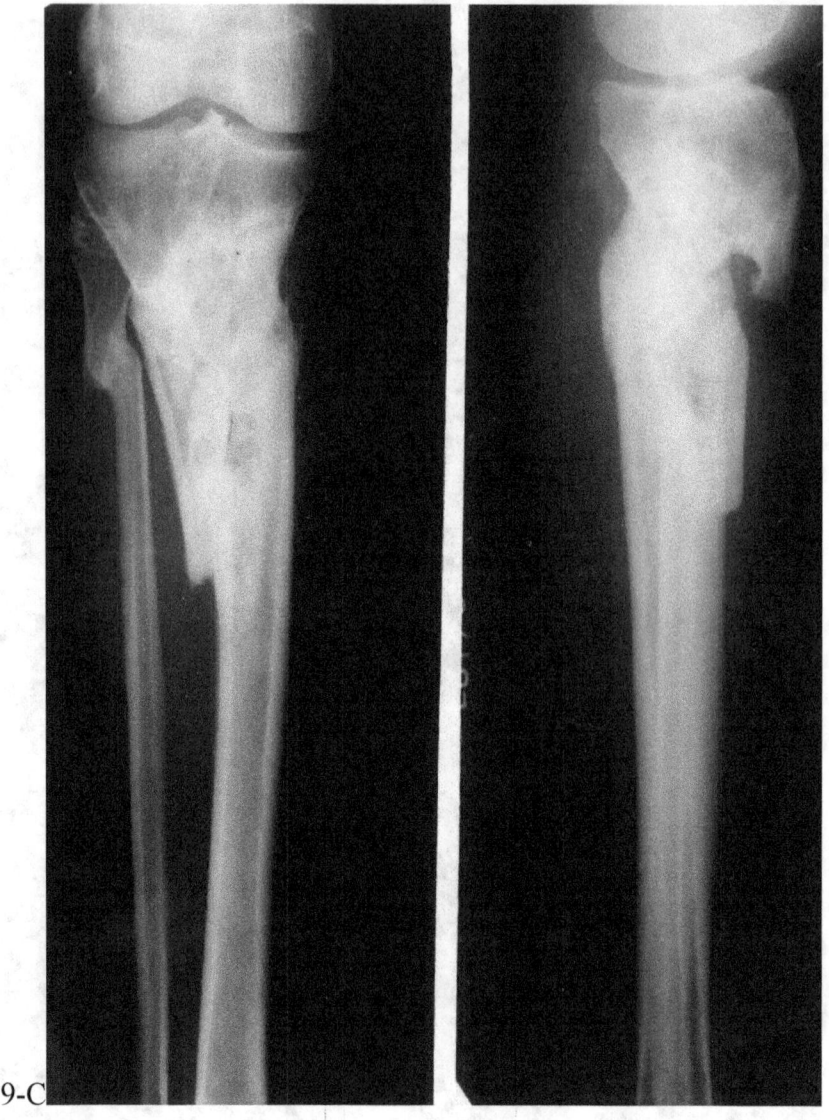

59-C

Figure 59-C   Antero-posterior and lateral roentgenograms upon completion of healing.

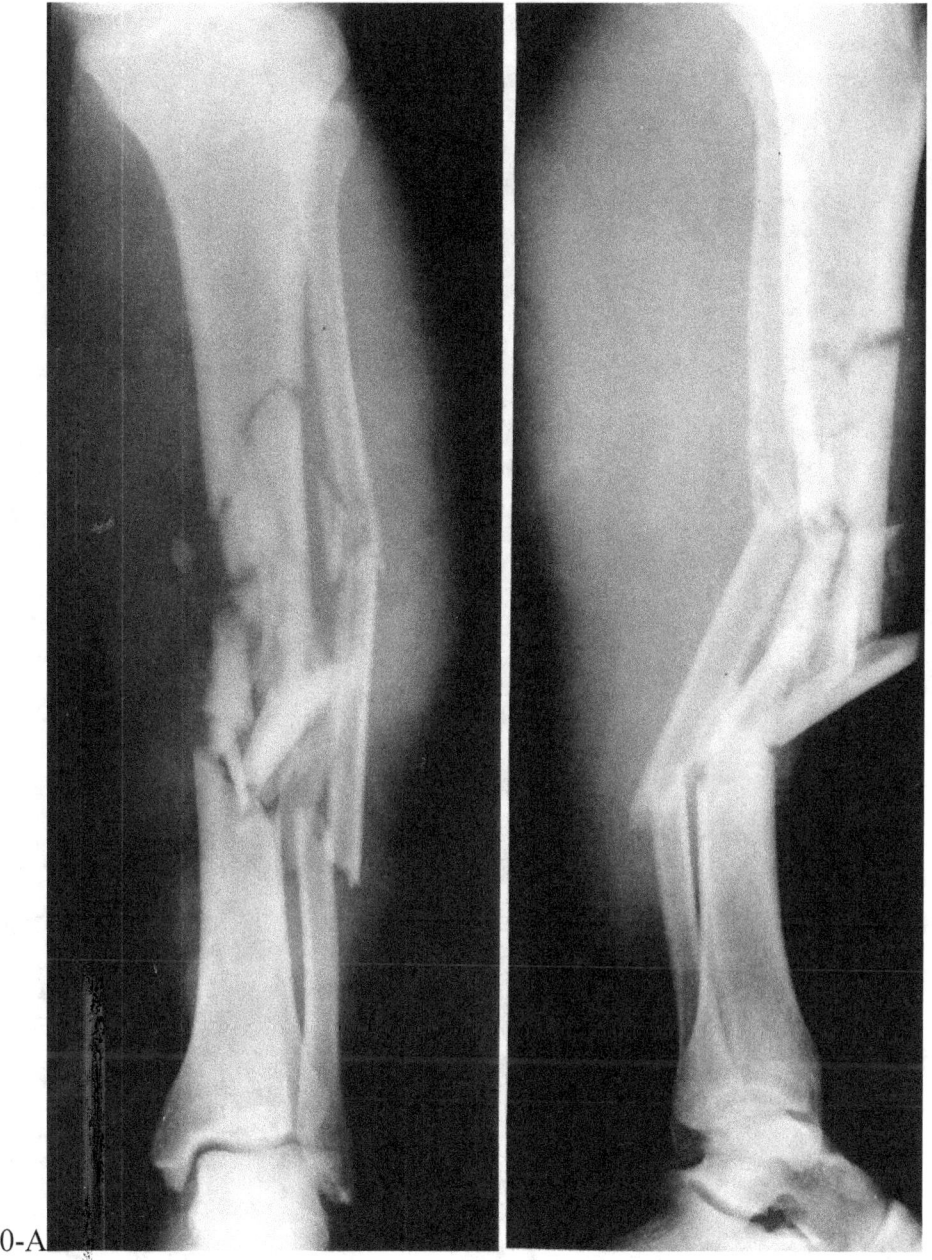

60-A

Figure 60-A   Antero-posterior and lateral roentgenograms of an open comminuted fracture of
the tibia and fibula

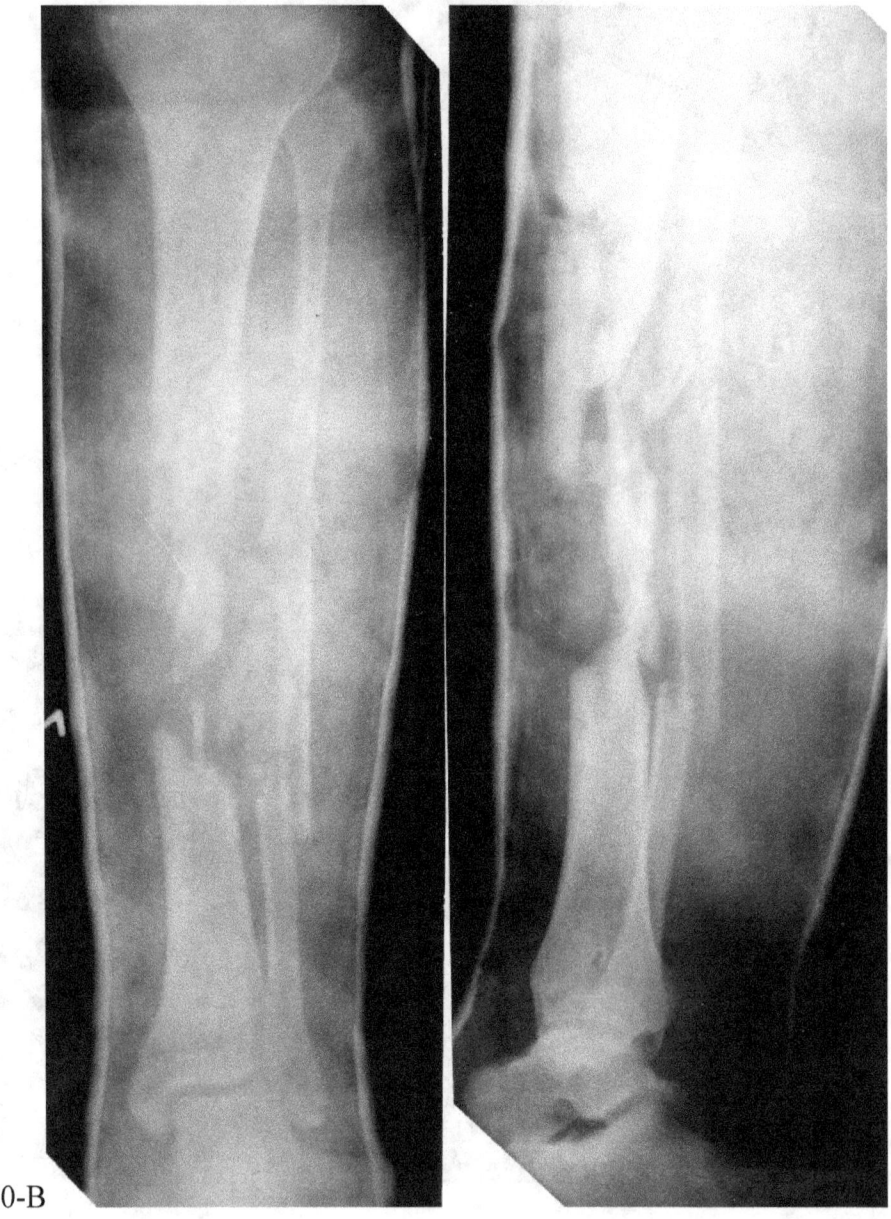

60-B

Figure 60-B   Antero-posterior roentgenograms following debridement and immobilization in a below-the-knee functional cast

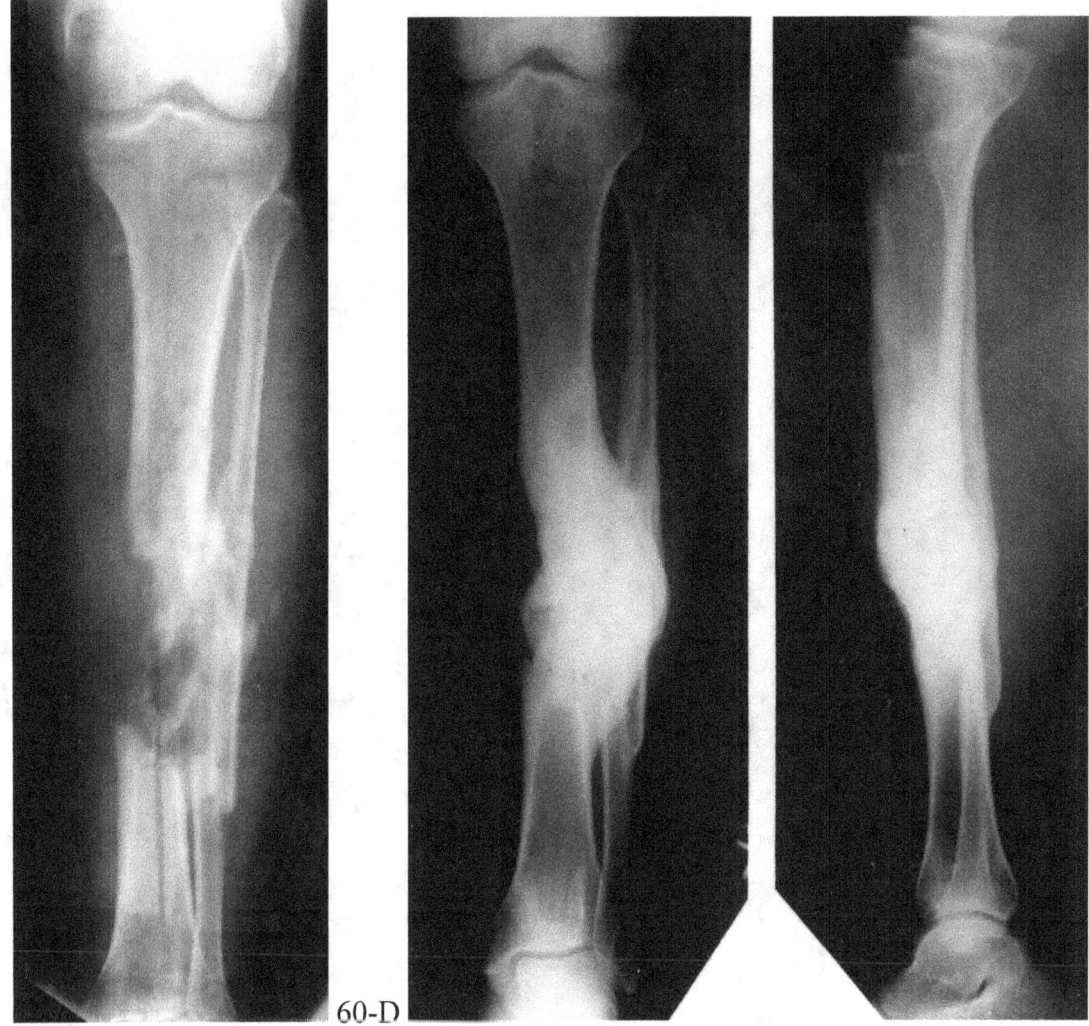

60-C   60-D

Figure 60-C   X-ray obtained through the below-the-knee functional Orthoplast brace

Figure 60-D   Antero-posterior and lateral roentgenograms upon completion of healing of the
fractures without apparent additional shortening of angulatory deformities.

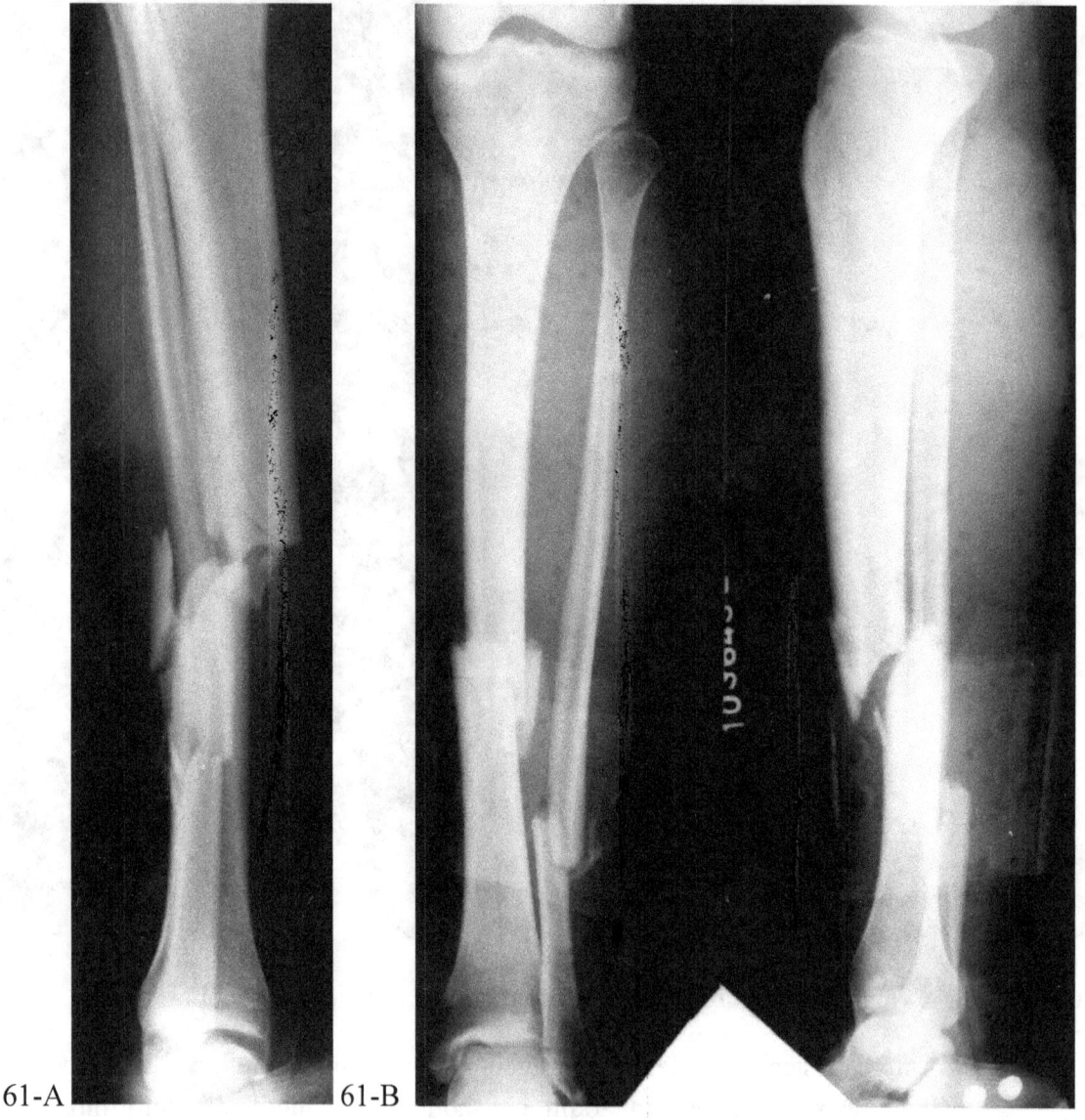

61-A   61-B

Figure 61-A   Closed fracture of the tibia and fibula.

Figure 61-B   Antero-posterior and lateral roentgenograms obtained through the below-the-knee functional Orthoplast brace.

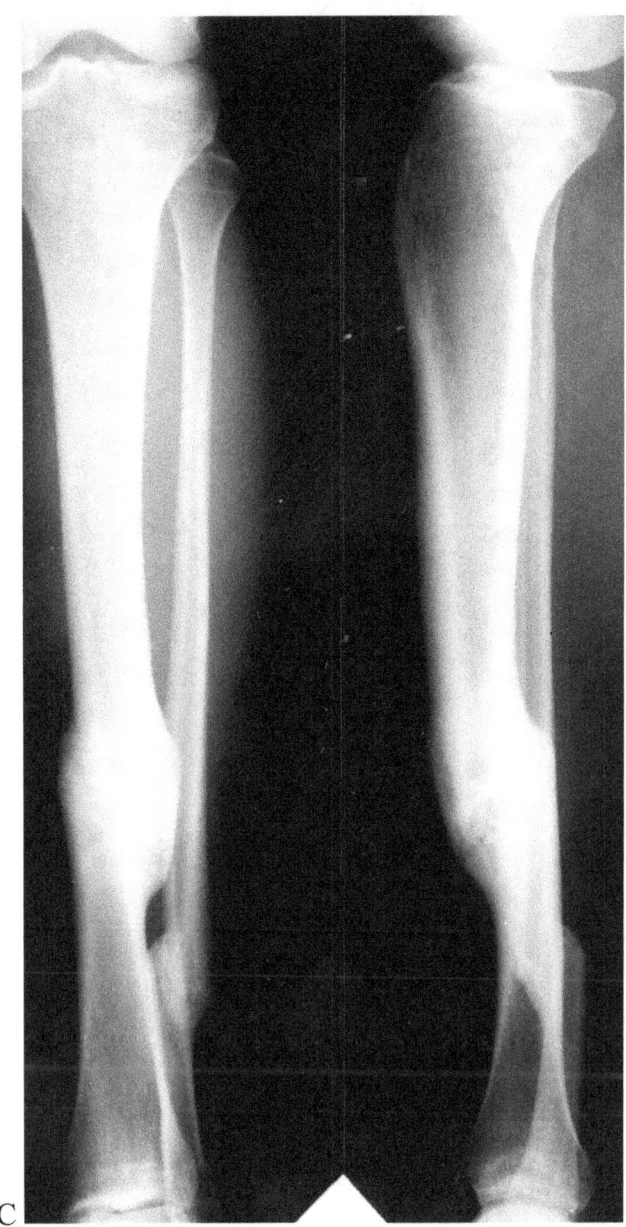

61-C

Figure 61-C   Roentgenograms obtained upon completion of healing illustrating maintenance
of length and alignment of the fragments.

Final Conclusions

linical experience with 532 cases of tibial fractures treated with braces that permit freedom of motion of all joints and early weight bearing ambulation with low morbidity and minimum residual disability, have suggested that rigid immobilization is not a prerequisite for fracture healing. The rather consistent early development of stable and abundant periosteal callus of good mechanical structure also suggests that the elastic motion at the fracture site, if a result of function, interferes minimally with rapid osteogenesis. We conducted experimental studies in rats which supported this concept by the fact that animals with fractured femurs healed sooner if permitted to ambulate without external fixation when compared with those immobilized in plaster casts. Mechanical properties of the healed bone in the functional animals exhibited greater strength and toughness. The microstructural appearance of the fracture callus also demonstrated better arrangement of materials and a larger mass of osteocartilagenous tissue which was responsible for the higher stiffness of the whole bone structure. It appears that the internal architecture of the healing bone is determined by external forces to which it is subjected and that therefore the subjection of healing fractures to external stimuli encourages the formation of a callus which assumes the mechanical characteristics of mature whole bone at an earlier date.

The interosseous membrane probably experiences different degrees and types of damage proportional to the type of injury and initial displacement of fragments. It assumes a major role in determining the initial and subsequent behavior of the fracture by providing an effective stabilizing mechanism.

The soft tissues of the lower extremity in conjunction with the below-the-knee functional brace through a hydraulic mechanism play an important role in maintaining the length of the fractured extremity. They also prevent further shortening by providing a support about which fragments can displace elastically allowing a stable motion to occur at the fracture site.

The shortening experienced by tibial fracture at the time of the original injury, if not artificially corrected, remains essentially unchanged. The degree of shortening is proportional to the severity of the soft tissue damage. Most closed tibial fractures with associated fibular fractures experience initially no more than one quarter of an inch of shortening. A healed extremity with even twice that degree of shortening should not result in a detectable limp or be expected to created a permanent functional impairment.

The functional fracture brace that we developed and have used during the past ten years, in conjunction with the soft tissues, provides the necessary stability required to prevent rotational, angulatory and longitudinal deformities. The high anterior and condylar extension of the brace proximally are important for controlling rotational stability. The foot-piece primarily stabilizes against angulatory deformities and secondarily against rotational deformities and clinically has proven to be important in holding the Orthoplast sleeve in place on the leg. The fit of the cylindrical portion of the brace is instrumental in containing the soft tissues and thus preventing shortening and angulatory deformities. Mobilization of the joints above and below the fracture site does not jeopardize mechanical stability of the fragments nor good fracture healing.

It appears that the near physiologic environment in which the limbs are placed throughout the entire healing process, provides a desirable milieu which is conducive to prompt and uninterrupted osteogenesis. The early mobilization of all joints of the injured extremity, the intermittent muscle contraction and weight-bearing, prevent atrophy, osteoporosis and joint stiffness. This functional method has demonstrated significant advantages in regard to the return of patients to the pre-fracture ambulatory status and the elimination of the need for rehabilitation of joints and muscles upon completion of healing. Oftentimes patients have been able to return to employment and productive roles in society while still under active treatment.

Through this method of treatment, osteomyelitis from open surgery has been virtually eliminated, the problems of delayed unions and non-unions almost completely resolved and healing time reduced. Clinical and laboratory experiments discussed in this paper have provided us with a better understanding of the behavior of the fractures and served as a stimulus for the development of similar functional methods of treatment for fractures of other long bones of the appendicular skeleton In varying degrees, the application of these principles to the treatment of fractures of the femoral shaft 53, the forearm 57, the wrist 56, the humerus and the tibial condyles 30 have been satisfactory.

REFERENCES

1. Abendschein, W. and Hyatt, G.: Ultrasonics and Selected, Physical Properties of Bone. Clin. Orthop., Q9; 294, 1970.

2. Abrahms, M.: Mechanical Behavior of Tendon in Vitro – A Preliminary Report. ~Med. Biol. Engr., §;433, 1967.

3. Allgower, M. et al: Clinical Experience with-a New Compression Plate "DCP" in Cortical Bone Healing. Acta Orthop. Scand. Suppl., 12§:43, 1969.

4. Anderson, L.D.: Compression Plate Fixation and the Effect ; of Different Types of Internal Fixation on Fracture Healing. J. Bone and Joint Surg., 47-A:191, 1965.

5. Boyd, H.B.: Delayed and Nonunion of Fractures. Campbell's Operative Ortho., Fifth Edition, Volume 1, St. Louis, The C.V..Mosby Companry, 1971.

6. Brooks, Murray: The Blood Supply of Bone. London, Butterworth and Company, 1971.

7. Charnley, J.: The Closed Treatment of Common Fractures. Third Edition, Williams & Wilkens Company, 1961.

8. Currey, J.D.: The Mechanical Properties of Bone, Clinical Orth and Rel. Res., 73:2`0, 1970

9. Diamant, J. et al.: Collagen; Utrastructure and its Relation to Mechanical Properties as a Function of Aging. Proc. R. Soc London. B180: 293, 1972.

10. Eggers, G.W.N.: Effect of Contact Compression on Osteogenesis Chapter 18, Instructional Course Lecture, A.A.O.S., 1952.

11. Indications and Operative Technique for Open Reduction and Internal Fixation of Fractures of the Shaft of the Tibia and Fibula. Surg. Clin. of N.A., §1:1515, 1961.

12. Fitts, W.T., et al: The Effect of Intramedullary Nailing on the Healing of Fractures. Surg. Gyn. & Ob., 89; 609, 1949.

13. Flint, M.: Interrelationships of Mucopolysaccharide and Collagen in Connective Tissue Remodeling. J. Embryol, Exp. Morph Z;481, 1972.

14. Friedenberg, Z. B. and French, G.: Effects of Known Compression Forces on Fracture Healing; Surg. Gyn. & 0b.,9§;743, 1952

15. Getser, M. and Trueta, J.: Muscle Action, Bone Rarefaction & Bone Formation. J. Bone and Joint Surg., 40-B: 274, 1958.

16. Gillespie, J.: The Nature of Bone Changes Associated with Nerve Injuries and Disuse. J. Bone and Joint Surg., 36-B: 464, 1954.

17. Gothman, L.: Arterial Changes in Experimental Fractures of the Rabbit's Tibia Treated with Intramedullary Nailing. Acta. Orth. Scand., 1202289, 1960.

18. _____: Arterial Changes in Experimental Fractures of the Monkey s Tibia Treated with Intramedullary Nailing. Acta. Orth. Scand., 121;56, 1961.

19. _____ : Local Arterial Changes Associated with Diastasis in Experimental Fractures of the Rabbit's Tibia Treated with Intramedullary Nailing. Acta. Chir. Scand., 123:104, 1962.

20. Grant, J.: Grant's Atlas of Anatomy. Fifth Edition, Williams and Nilkens Company, 1962.

21. Gray, H.: Gray's Anatomy. Twenty-eighth Edition. Edited by C. Goss, Les and Febiger, 1968.

22. Hart, C.R., Hale, M.S. and Burkhalter,N.: Ambulatory Electromyographic Studies in Patients with Tibial Fractures in Long Leg Casts and Below the Knee Casts. J. Trauma, 12:223,1972.

23. Hicks, J.H.: External Splintage as.a Cause of Movement in Fractures. Lancet, 1;667, 1960.

24. Holden, C.E.A.: The Role of Blood Supply to Soft Tissues in the Healing of Diaphyseal Fractures. J. Bone and Joint Surg., 54-A1993, 1972.

25. Hults, A. and Olerud, S.: The Healing of Fractures in Denervated Limbs. J. Trauma, 5;571, 1965.

26. Jackson, R. et al.: Production of a Standard Experimental Fracture. Can. J. of Surg., 13;4.

27. Jackson, R.N. and McNab, I: Fractures of the Shaft of the Tibia. Am. J. Surg., 97;543, 1959.

28. Karlstrom, G. and Olerud, S.: Fractures of the Tibial Shaft -A Critical Evaluation of Treatment Alternatives. Clin. Orthop. & Rel. Res., 105;82, 1974.

29. Kernek, C.B. and Wray, J.B.: Cellular Proliferation in the Formation of Fracture Callus in the Rat Tibia. Orthop., 91: 197, 1973.

30. Kinman, P. et al.:' Experimental Tibial Condylar Fractures. J. Bone and Joint Surg., 57-A:576, 1975. T

31. Lambert, K.L.: The weight—bearing Function of the Fibula -J. Bone and Joint Surg., 53 A:507, 1971

32. Lindholm, R.B. et al.: 'Effect of Forced Interfragmental Movements on Healing of Tibial Fractures in Rats. Acta. Orth. Scand., 402721, 1970.

33. Lindsay, M.K. and Howes, E.L.: The Breaking Strength of Healing Fractures. J. Bone and Joint Surg., 13;491, 1931.

34. Lippert, F.G. and Hirsch, C.: The Three-Dimensional Measurement of Tibial Fracture Motion by Photogrammetry. Clin. Orthop. and Rel. Res., l05:130, 1974.

35. Macnab, 1.: The Role of Periosteal Blood Supply in the Healing of Fractures of the Tibia. _Clin. Orthop. and Rel. Res., 105;27, 1974.

36. McElhaney, J. et al.: Effect of Embalming on the Mechanical- Properties of Beef Bone. J. App. Physio., 19-6:1234, 1964.

37. Milner, J. C. and Rhinelander, F.W.: Compression Fixation in Primary Bone Healing. Surg. Forum, 19;453, 1968,

38. Mindell, E., Rodband, 5., Kwasman, B.: Chondrogenesis in Bone Repair. Clin. Orthop. and Rel. Res., Z9:187, 1971.

39. Olerud, S. and Danckwardt—Lilliestrom: Fracture Healing in Compression Osteosynthesis in the Dog. Instructional Course Lecture, A.A.0.S., J. Bone and Joint Surg., 50-B:844,'1968.

40. _____:Fracture Healing in Compression Osteosynthesis. Rcta. Orthop. Scand. Supp.,137,.1971.

41. Paradis, G.R. and Keny, P.J.: Blood Flow and Mineral Deposition in Canine Tibial Fractures, J. Bone and Joint Surg., 57-A:220, 1975.

42. Phemister, D.B.: Biologic Principles in the Healing of Fractures and-Their Bearing on Treatment. Ann. of Surg.,133;433, 1951.

43. Piekarski, K., Wiley, A.M., Bartels, J.E.: The Effect of Delayed Internal Fixation on Fracture Healing. Acta. Orthop Scand., 4Q:543, 1969.

44. Reilly, D.T. and Burstein, A.: The Mechanical Properties of Cortical Bone. J. Bone and Joint Surg., 56 A:1001, 1974.

45. Reynolds, F.C. and Key, J.A.: Fracture Healing After Fixation with Standard Plates,LLontact Splints and Medullary Nails. J. Bone and Joint Surg., 36-A:577, 1954.

46. Rhinelander, F.W. et al.:'Microangiography in Bone Healing III: Osteotomies with Internal Fixation. J. Bone and Joint Surg., 49-A:1006, 1967.

47. Rhinelander, F.w. and Baragry, R.: Microangiography in Bone Healing, II Displaced Closed Fractures. J. Bone and Joint Surg., 50-A1643, 1968.

48. Rhinelander, F.W.: Tibial Blood Supply in Relation to Fracture Healing. Clin. Orthop. and Rel. Res., 105:34, 1974.

49. Robin, G., Levin, S., and Wilin, A.: Dynamic Breakability of Bone. Clin. Orthop. and Rel. Res., 69;289, 1970.

50. Sarmiento, A.: A Functional Below-the-Knee Cast for Tibial Fractures. J. Bone and Joint Surg., 49-A:855, 1967.

51. Sarmiento, A., and Sinclair, H.F.: Application of Prosthetic-Orthotic Principles to Orthopaedics. Art. Limbs, 2:2, 1967.

52. Sarmiento, A.: A Functional Below-the-Knee Brace for Tibial Fractures. J. Bone and Joint Surg., 52-A2295, 1970.

53. Sarmiento, A. : Functional Bracing of Tibial and Femoral Shaft Fractures. Clin. Orthop. and Rel. Res., B252, 1974.

54. Sarmiento, A., Latta, L., Zilioli, A. and Sinclair, H.F.: The Role of Soft Tissues in Stabilization of Tibial Fractures. Clin. Orthop. and Rel. Res., 105;116, 1974.

55. Sarmiento, A.: Functional Bracing of Tibial Fractures. Clin. Orthop. and Rel. Res., , 1974.

56. Sarmiento, A., et al.: Colles' Fractures. VJ. Bone and Joint Surg., 57-A:311, 1975.

57. Sarmiento, A., Cooper, J.S. and Sinclair, N.F.: Forearm Fractures. Early Functional Bracing - A Preliminary Report. J. Bone and Joint Surg., 57-A:297, 1975.

58. Semb, Helge: Experimental Limb Disuse in Bone Blood Flow. Acta Orthop. Scand., 40;552, 1969.

59. Shaw, J., and Bassett, C.A.L.: The Effects of Varying Oxygen -Concentration on Osteogenesis and Embryonic-Cartilage in Vitro; J..Bone and Joint Surg., 49-A:74, 1967.

60. Stillwell, G. K.: The1Law of Laplace - Some Clinical Applications. Mayo Clin. Proc., 48;863, 1973.

61. Trueta, J.: The Role of Vessels in 0steogenesis.¥ J. Bone and Joint Surg., 45-B:402, 1963.

62. Trueta, J. and Buhr, A.J.: The Vascular Contribution to Ostegenesis, V - The Vasculature Supplying the Epihyseal Cartilage in Rachitic Rats. J. Bone and Joint Surg., 45-B:572, 1963.

63. Trueta,J.: Blood Supply and Rate of Healing of Tibial Fractures. Clin. Orthop. and Rel. Res., 1Q§;11, 1974.

64. Ududa, K. and Prasad, G.: Chemical and Histological Studies on Organic Constituents in Fracture Repair in Rats. J. Bone and Joint Surg., 45-B:770, 1963.

65. Urist, M. and McLean, F.: Calcification in the Callus in Healing Fractures in Normal Rats. J. Bone and Joint Surg., 23;1, 1941.

66. Urist, M.R. et al.:s Pathogenesis and Treatment of Delayed Union and Non-Union. J. Bone and Joint Surg., 36-A:931, 1954.

67. Varma, B.P. and Mehta, S.H.: Fracture Healing with Intra-medullary Nail Fixation of the Long Bones. Acta. Orthop. Scand 38;419, 1967.

68. Viidik, A.: Simultaneous Mechanical and Light Microscopic Studies of Collagen Fibers. Z. Anat. Entwickll Gesch., B-136: 204, 1972.

69. Watson-Jones, R. and Roberts, Calcification, Decalcification and Ossification, Par Brit. J. Rad., Z;321, 1934.

70. Weir, J.B. De V., Bell, G.H., Chambers, J.w.: The Strength and Elasticity of Bone in Rats on a Rachitogenic Diet. J. Bone and Joint Surg., 31-B:444, 1949. V

71. Nolinsky, H. and Glagov, S.: Structural Basis for the Static Mechanical Properties of the Aortic Media. Circ. Res.,_Q4: 400, 1964.

72. Yamagiski, M. and Uoshimura, Y.: The Biomechanics of Fracture Healing. J. Bone and Joint Surg., 37-A:1035, 1955.

PERSONNEL AND AGENCIES INVOLVED IN THIS PROJECT

In addition to the authors, the following people contributed to the work included in this paper. The fracture healing studies performed in rats were directed by John Schaeffer, M.D., with engineering assistance by Linda Beckerman, M.S.

Studies of the load bearing capacity of the functional below-theknee fracture brace were performed by Robert Posival, M.S., with the assistance of William Sinclair, C.P.0.

Studies of the behavior of tibial fractures in the laboratory and clinical practice, the role of soft tissues in stabilization of fractures and the design of the prefabricated fracture braces received contributions from: Armand Zilioli, M.D., William Sinclair, C.P.O., and James Ruopp, M.S.

In addition to the facilities of the University of Miami School of Medicine for animal surgery, microscopy and fabrication of fixtures, braces, etc., the Veterans Administration Hospital biomedical engineering laboratory was used for mechanical testing studies. The School of Engineering at the University of Miami made available to us their mechanical testing laboratory. Histology preparations were done by the Department of Pathology at Jackson Memorial Hospital.

www.ingramcontent.com/pod-product-compliance
Lightning Source LLC
Chambersburg PA
CBHW081128170526
45165CB00008B/2586